FOLDING KNIVES
CARRY & DEPLOYMENT

Steve Tarani

EMPIRE Books
P.O. Box 491788, Los Angeles, CA 90049

www.empirebooks.net

Disclaimer

Please note that the author and publisher of this book are NOT RESPONSIBLE in any manner whatsoever for any injury that may result from practicing the techniques and/or following the instructions given within. Since the physical activities described herein may be too strenuous in nature for some readers to engage in safely, it is essential that a physician be consulted prior to training.

Published in 2007 by Empire Books.
Copyright © 2007 by Empire Books.

All rights reserved. No part of this publication may be reproduced or utilized in any form or by any means, electronic or mechanical, including photocopying, recording, or by any information storage and retrieval system, without prior written permission from Empire Books.

Library of Congress: 2007040579
ISBN-13: 978-1-949753-10-3

Library of Congress Cataloging-in-Publication Data
Tarani, Steve
  Folding knives : carry and deployment / by Steve Tarani.
    p. cm.
  Includes bibliographical references and index.
  1. Pocket knives. I. Title.
 TS380.T37 2007
 355.5'48--dc22
          2007040579

Empire Books
P.O. Box 491788
Los Angeles, CA 90049
(818) 767-7900

First edition
07 06 05 04 03 02 01 00 99 98 97 1 3 5 7 9 10 8 6 4 2
Printed in the United States of America.

Editor: John Spezzano
Cover & Interior Design
& Production: Mario M. Rodriguez, MMR Design Solutions

FOLDING KNIVES

CARRY & DEPLOYMENT

Steve Tarani

EMPIRE Books
P.O. Box 491788, Los Angeles, CA 90049

www.empirebooks.net

Folding Knives: Carry & Deployment

Table of Contents

Foreword . . . . .page 6
Acknowledgements . . . . .page 11
Author's Note . . . . .page 12
Background . . . . .page 13
Scope of Study . . . . .page 16
Model Policy . . . . .page 18
Concepts and Issues . . . . .page 21

Part I Cro-Magnons to Cryogenics . . . . . Page 24

Folding Knives – A Brief History . . . . .page 25
Categories of Knives . . . . .page 27
Fixed and Folding
Fixed Blades
Folding Blades
    Mechanical
    Unassisted
    Gravity Assist
    Spring Assist
    Autos
    Semi-autos
Metallurgy and Blade Geometry . . . . .page 32

Part II Parts and Selection . . . . Page 40

Folding Knife Parts . . . . .page 42
Parts of the Blade . . . . .page 42
Blade Shape . . . . .page 42
Straight Blade
Curved Blade
Blade Steel . . . . .page 44
CPM S30V
5160.52100, L-6, A-2, D-2, M-2
440C, 145CM, ATS-34, AEBL, 420
VASCO, Damascus, Stellite 6-K, Titanium
HRC 57-58, ATS-55, CPM-10V, N690

Contents

Blade Points . . . . .page 51
Spear Point
Drop Point
Clip Point
Angled Point (aka "Tanto")
Sheep's Foot
Hawk Bill

Blade Edges . . . . .page 55
Single Edged
Double Edged
Plain Edge
Blade and Edge Grinds
Blade Grinds
    Flat Grind
    Hollow Grind
    Convex grind

Edge Grinds . . . . .page 57
    Chisel Bevel
    Single Bevel
    Double Bevel
Serrations
    Full Serration
    Half Serration

Blade Finishes . . . . .page 63
Bead Blasting Finish
Coating Finish
Polished Finish
Satin Finish
Camouflage

Parts of the Handle . . . . .page 66
Friction Radius
Finger Cutout
Carry Clip
    Anchor Holes
Lanyard Hole
Handle Materials

Opening Mechanisms . . . . .page 70
"T" Stops
Pins
Posts
Wave
Holes

Ovals
Indents
Groves

Locking Mechanisms . . . . .page 80
Latch Lock
Clasp Lock
Back Lock
Front Lock
Liner Lock
Frame Lock
Axis Lock
Nitrous Lock
Rolling Lock
Modified Locking Liner
Mono Lock
Automatic Lock

Folding Knife Selection . . . . .page 90

Preventative Maintenance (PM) . . . . .page 94
Sharpening
Lubrication
Storage

Part III Operational Proficiency . . . . . . Page 96

Operational Skills . . . . .page 97
Training Methodology
Motor Skills

Folding Knife Carry . . . . .page 98
Accessibility
Viable Locations

Locking and Unlocking . . . . .page 105
Varying Methods

Safe Open . . . . .page 106
Locking
Ensuring a Stable Lock
Safe Close
Unlocking
Ensuring a Safe Close

Contents

Getting a Grip ....page 116
Determining Your Grip
Specific Grips
Optimal Grip Open
Optimal Grip Closed

Rapid Deployment ....page 122
Presenting the Folding Knife
Deployment Training Drills
    Drill One
    Drill Two
    Drill Three
    Drill Four
The Doctor's Wife
Perishable Skills

Position and Balance ....page 132
ARC in your AO
Stable Working Platform

Safe Handling ....page 134
Engaging the Blade Edge
    Rigid and Flexible Cuts
    Double Edge
    Single Edge
    Plain Edge
    Serrated Edge
    Combo Edge
Engaging the Blade Point
    Drop Point
    Angle Point
    Coefficient of Friction

Conventional Positions ....page 144
Standing
Kneeling
Seated

Unconventional Positions ....page 148
Prone
Supine
Asymmetric

Conclusion ....page 151

About the Author ....page 152

Foreword

A strong interest in knives started very early in my life. My father was a shoe repairman and my grandfather was a butcher by trade, who had a passion for hunting. Consequently, I was around knives and other edged tools from a very early age. Both men taught me to use and respect knives properly. I didn't know it then, but I know now that this was the base line for my love of knives. They trusted me enough to let me save my own money and purchase some of my first knives from the local hardware and department stores. The "pocket knives" I began to carry in my pocket became tremendous tools for a kid. In fact I took one to show and tell in one of my early elementary school years and got in trouble for doing it! No knives at school!

During my high school years, and for a very short time afterward, I worked in the grocery business as a clerk. My "Buck knife" was always in the case on my belt, and I used it daily for a variety of tasks, from cutting cardboard to scraping the sticky price tags off cans that had to be re-priced. This is when I began to realize that for many chores, the knife was not the perfect tool for the job at hand, but is was usually the one you always had with you. Adapt, overcome, and improvise! Anyone that has carried a knife for any amount of time has done all three of these.

My law enforcement career began in 1979. I was very young, and wanted to be just like the seasoned veterans. I was and will always be an equipment junkie, and so when I saw the veterans carrying their Buck knives, I quickly joined their club. I'm not sure that the administration even knew that we had them. There was certainly no written "police duty knife" policy, which many agencies including mine have adopted today, but they were always on our belts. If you were really cool, you would bevel the brass ends of the grip to customize your Buck. I'm sure this is not recommended by the company, but it gave a sleek sleeve rookie some veteran machismo. The pocket clip was years down the road, and these knifes were heavy so they were routinely carried on the duty belt in a basket weaved magazine pouch that matched the other leather gear on your belt. Later a knife pouch was actually made for the folding knife, but the magazine pouch always worked fine, and the general public never knew the folding knife was there.

Then it happened, the Spyderco Police Model hit the market, we all went crazy! Flat, sharp, serrated edge, a thumbhole for one-handed opening, and a strong lock up! You could it carry on and off duty, it was easy to maintain and was as sharp as a razor. This in my opinion was the nucleus that started the Tactical Knife era of today, which has led to the development of some tremendous folding knives for use by public safety personnel and the informed private citizen.

Currently, and in the past twenty-five years in law enforcement, I have used my folding knife for a variety of tasks; poking and prodding hundreds of crime scenes, collecting evidence, cutting seatbelts, cutting away airbags that had been deployed at the

Foreword

scene of car accidents, cutting window molding along a vehicle's window to make space for my "Slim Jim" when a panicked mother accidentally locked her infant child and her cars keys in the car. I have cut clothing off a critically ill subject with a folding knife because it was the only tool I had to do the job, in an attempt to save his life. I don't recommend this, but I made the decision based on the circumstances presented to me.

A folding knife was used by a member of my department to cut a garden hose, which was then attached to an oxygen bottle, and lowered approximately 20 feet into an abandoned well shaft to provide oxygen to a small child who had fallen down the shaft and was trapped for hours until successful rescue, and yes, the folding knife has been used to cut the rope of people attempting suicide. These are quick, sometimes split-second decisions where the folding knife was used to solve a problem. Knives, in my opinion, are problem solving tools.

Both on and off duty we carried it on our person where it was accessible. This accessibility is what made is such a viable tool. We, as law enforcement professionals, have all pried on windows and doors with our folding knife of course, with the legal reason to do so. And yes we have cut more then one sandwich in half, and forced open our share of soda cans with them too. Today I carry two folders when on duty. One is a full size Tactical Folder that I use for heavy chores, and a small folder that I use for mundane tasks.

Since September 11, 2001, the Public Safety word of choice to describe Law Enforcement, Fire Fighters, Paramedics, EMT's, health officials and others is "First Responders." I cannot over state the need for a folding knife as a piece of essential equipment for these professionals. We, as law enforcement professionals, have learned from first-hand experience, that you survive and rescue people initially with what equipment you have available on your person. Such equipment should include a folding knife.

Some may take exception to my opinion, and debate the need for Public Safety Personnel to carry folding knives. I can only say that I lose no sleep over it. If you trust an officer enough to carry a variety of firearms to employ deadly force when legally and morally justified, and drive a motor vehicle at high speed in all conditions, then you certainly can trust them to carry a folding knife. In fact Steve Tarani, the author of this book wrote my department's policy on the carry and use of the folding knife by members of my department that I adopted as our SOP. Steve then delivered a very basic safe knife handling class for all personnel that was approved by the California Commission of Peace Officer Standards. Simple, efficient, and important training.

In the shadow of 9/11 the private citizen, who may read this book should also know that they have a role in their personal survival, be it an auto accident, a medical emergency, an outdoor survival scenario, or terrorist attack. My wife and daughter both carry a tactical folding knife, multi tool and flashlight in their purses, every-

Folding Knives: Carry & Deployment

where they go. They both complained initially, but now I see them using these tools occasionally to solve some of life's little problems.

Your choices available for folding knife selection are endless. But I think there are a few simple rules to consider. Expect to spend at minimum a reasonable amount of money. There is a folding knife made by respected companies in many price ranges, so you will be able to find one that will work for you. Be conscious of quality. Don't buy cheap imitations of the large recognized companies and custom makers in the industry. Here are some of the selection criteria I have used to choose the make and models that I carry.

- ▼ A good quality high-grade steel blade of 3.5 to 4 inches in length with a straight edge, and if possible a tanto point. I work and carry the knives primarily in an urban environment, and I find that this combination works best for me. But run some simple tests for yourself to see if the blade will cut and penetrate what you want it to. My knives will cut plastic, heavy clothing, and the tanto (angled) tip can penetrate dry wall.
- ▼ The locking mechanism must be extremely strong, or you will get hurt when the blade collapses on your hand. You should also make sure that you could release the lock with one hand if necessary, and in the dark.
- ▼ The grip surface should be rough. I like G10 or another material that is strong, and gives me a rough griping surface even when my hands are wet or cold. The grip should be impervious to fluids.

Maintenance of your personal protective equipment has always been an issue with me. You would not drive your car without engine oil, or use your laptop without virus protection, but many law enforcement officers, young and old, that I am familiar with refuse to maintain their firearms, folding knives, handcuffs, flashlights, etc. These are all mechanical devices that will fail to function without proper preventive maintenance (PM). The folding knife needs to be kept clean and lubricated to the manufacturer's recommendations. Do not try to take a folding knife apart, if it needs that amount of maintenance; return it to the manufacturer's customer service department.

As I write this, there are thousands of U.S. military personnel and other First Responders who willingly place themselves in harm's way. In their operational tool kits are quality tactical folding knives that have saved lives in a variety of ways. This manuscript will provide you, the law enforcement officer and informed citizen alike, a wealth of information to appropriately select, carry and utilize this vital piece of safety equipment in our ever-changing world.

Ron Langford
Chief of Police
Del Rey Oaks Police Department
California, USA

Acknowledgments

Special thanks to Chief Ron Langford (Del Rey Oaks Police Department, California) and Duane Dwyer of Strider Knives, Inc., for their contributions to this manuscript, Paul "Ted" Bubba Grybow and Tim Egberts, for assisting with the illustrations, Barry R. Shreiar for his continued support and generosity without whom this project would truly never have been completed, the folks over at CPM, Benchmade, SpyderCo, Emerson, Buck Knives, Camillus, Blade-Tech Industries and 5.11 Tactical for their input to these materials and John Spezzano for his most generous offer of time and effort in editing and a long-overdue resounding "thank you" to Sheriff Ken Campbell, Mark Babyak and Tom Petrowski for their contributions throughout the early development phases of this research material. Additionally, it is with tremendous respect and honor that I acknowledge Dan Inosanto, Leovigildo M. Giron and Edgar G. Sulite for their life-long inspiration, training and guidance.

Author's Note

In today's modern age of metallurgy, good quality knives are readily available. Specifically designed for the rigors of law enforcement and military work, folding knives are more than qualified for civilian use in commercial industry, construction, camping, hunting, fishing and other outdoor and indoor applications. The folding knife is a problem solving tool and as such is deserving of further study.

As a direct result of literally hundreds of questions fielded over the many years throughout professional training courses which I have had the privilege to deliver to the US military as well as federal, state and municipal law enforcement employees referencing folding knives, this manuscript has been designed to provide well-researched answers to anything you ever wanted to know about folding knives regarding selection, carry and usage.

If you happen to be a law enforcement administrator, police management staff or supervisory personnel and you're looking to utilize this book for support documentation (especially with regards to department / agency policy) then you're on target because that is exactly what this book is intended for – a solid reference manual to provide sound information. Please don't get jittery about the word "deployment" as part of the title. This book was specifically and painstakingly written devoid of any Defensive Tactics, martial arts, Karate, Kung Fu, self defense, officer survival techniques or otherwise. Nowhere in this manuscript will you find any reference whatsoever to a knife used as a weapon. You will find reference to application of the knife as a tool only.

Keep in mind that I too am a sworn officer and also as a professional contractor have fought the very same uphill battles for nearly a quarter of a century with scores of bosses and anti-knife administrators in trying to bring valuable information referencing folding knives to the forefront. Therefore any and all reference to the folding knife is purposefully and strictly limited to the selection, appropriate carry and usage of the folding knife as a utility tool only.

Included herein you will also find reference and some of the most detailed information available to date on Officer Knife Usage Policy as per the International Association of Chiefs of Police (IACP). You will also find a comprehensive and thorough Officer Knife Usage "Concepts and Issues" excerpt that was compiled with the aid of a very close associate of mine who, in the late '90's was General Counsel to a major US federal agency. In other words, the material comprising this manuscript is very heavily researched, includes direct contributions from leading industry manufacturers, meticulously reviewed by legal council (both federal and state), recommended standard policy per IACP and beta-tested in the field by military, law enforcement and related personnel at all levels for at least a decade.

You can rest assured that this extensively researched knife policy and policy-related knife usage materials are currently utilized by a number of federal, state, and municipal agencies. In fact, as the architect of both the IACP Model Policy for the carry, deployment and safe usage of knives by peace officer and the IACP Concepts and Issues Paper (policy support documentation) as well as the actual training Key (TK#576), I assure you the training materials and methodology herein coincide with IACP Policy, training and support documentation.

Background

In the days long before computers and video games (and at the risk of carbon-dating myself), I was a Cub Scout back in the late 1960's during which time the Camillus Knife Company produced a full line of official folding knives for the Boy Scouts of America (BSA). Camillus Knives (founded in the late 1800's in Camillus, New York), manufactured and shipped over 750,000 knives of varying styles to the allied forces during World War II, between 1942 and 1945. In 1947, Camillus began to produce folding knives for BSA and continued to do so until the closing of the factory in February of 2007. To this very day I can still remember that happy day of my youth when my fellow Cub Scouts and I were each issued a brand new official BSA Camillus folding knife. What really sticks in my mind was how cool it was that we each had our very own knife. There wasn't one member of our little band of newly ordained knife-owners without a beaming, ear to ear grin.

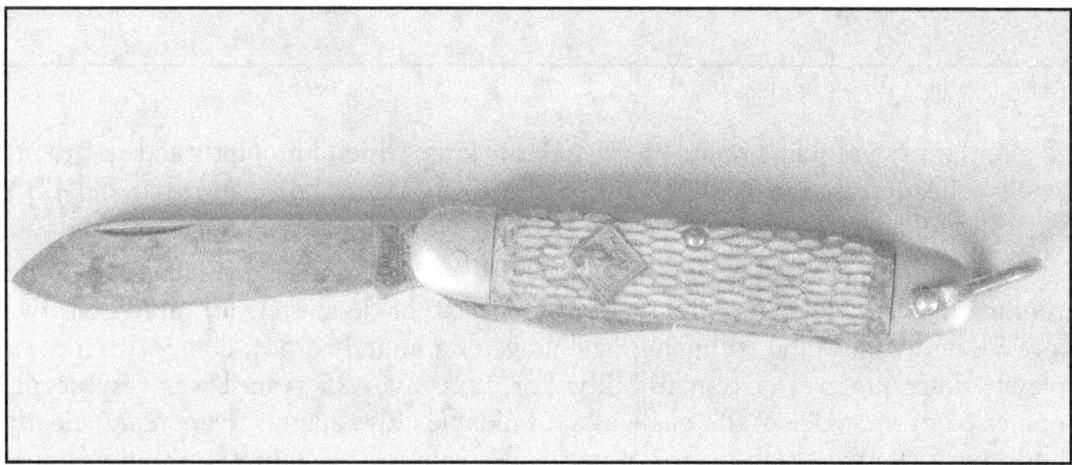

Original Camillus BSA folding knife (circa 1966)

The knife came with instructions on how to sharpen and, of course, from BSA leadership how to safely open, safely close and actually use the knife. We later utilized the blade for countless projects such as whittling, woodwork, camping, hiking and every other merit badge you could earn with the aid of that cool little folding knife. Another snapshot that remains etched in memory is that along with the issuing of that knife came the issuing of the responsibility of its use and care.

Later on I was employed as a construction worker and as such got plenty of experience with edged tools on the job. Working in construction I saw first hand how carelessness and irresponsible behavior with an edged tool (by irresponsible individuals) led to serious bodily injury. Determined to get through college with all ten fingers and ten toes, I made it a priority to not only pay attention to my handling but also the handling around me of my co-workers of edged tools. After a few paychecks I was able

Folding Knives: Carry & Deployment

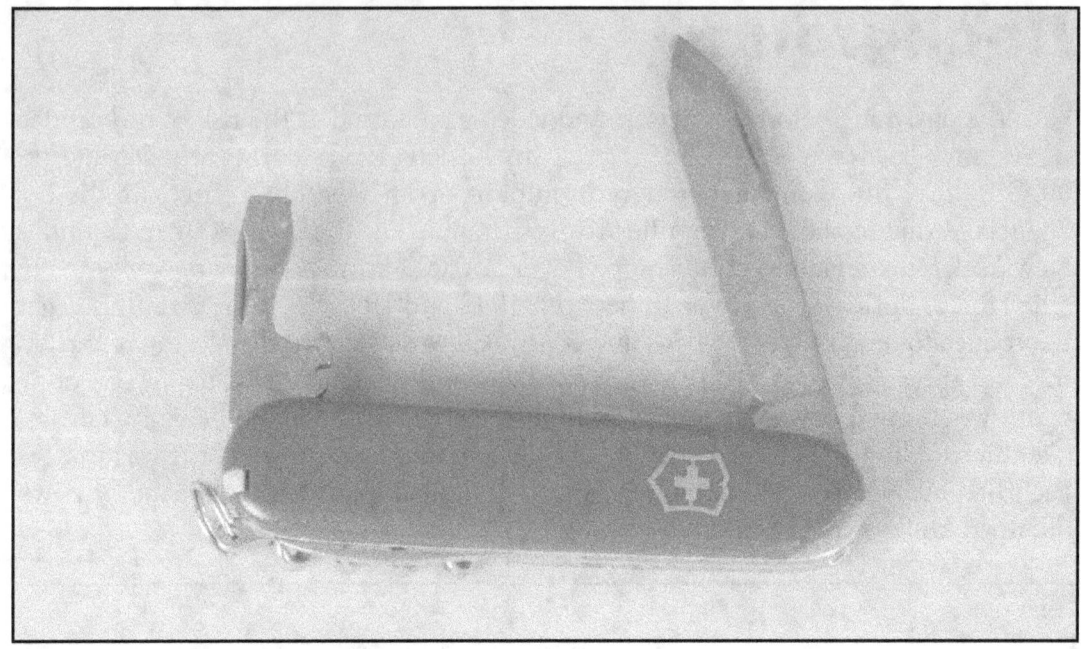
Swiss Army folding knife (circa 1980)

to afford my next folding knife – a Swiss Army knife which I promptly added it to my small collection of two folding knives: the Swiss Army knife and that old BSA Camillus folder.

Much later on as a devout student of Edged Weapons Defense (weapon arts) and training under Malaysian, Filipino and Indonesian blade masters for more than two and a half decades (and still going) the usage of a knife had now come with a completely different level of responsibility. For more than 25 years I was very deeply immersed in the usage of the blade as a formidable close-quarter weapon and clearly understood the very high degree of personal commitment and responsibility that is part and parcel of acquiring such knowledge.

Most alpha males (if you're reading this book, then there is a very good chance that you may be an alpha male) look at a knife the same way we look at a car or a gun – all alpha males are born knowing innately (akin to salmon knowing how to swim upstream) how to use a knife, shoot a gun and drive a car. Nonetheless, the intent of this book is to provide a comprehensive study of the selection, carry and usage of folding knives. Even way back in the days of the Cub Scouts of the late 1960's they didn't go into the kind of detail you will find here in these chapters.

As a sheriff's deputy, I noticed that although it wasn't officially issued by the department, most cops carried a knife. Officers and federal agents get training on guns, radios, driving (EVOC), baton and even on the usage of a flashlight (low-light small arms training). Most of the training is of course related to firearms because of the high liability, but much like the gun and the radio and the flashlight and the baton (all of which come with training), the knife (which does NOT come with any training –

hence the impetus for this book) is worn on the official uniform, utilized in the line of duty and in plain view of the public. Most departments won't even entertain the idea of maintaining a knife policy as it is less of a financial risk to place the onus of responsibility on the individual officer as opposed to burdening the department.

If something goes wrong – and it does on a regular basis (you would be surprised to know some of what really happens behind the scenes with cops and knives) – it usually gets swept under the rug as such incidents are completely overshadowed by gun-related incidents (the incidents most promoted by the media). In the rare event that a knife incident does make it to the media-level, then upper chain is rapid to respond with "well, uh, we uh didn't issue that uh, particular piece of equipment that was utilized by that officer and thus since it is an unauthorized piece of gear the department assumes no responsibility for the uh, activities of the officer utilizing said unauthorized gear and therefore this department cannot be held liable."

Scope of Study

Born from the most frequently asked questions of literally hundreds of students (probably in the thousands by now) as well as the requests from numerous departments, agencies and countless law enforcement, military and federal students since the late 1980's, this book is intended to provide a comprehensive overview of the selection, carry and usage of the modern folding knife.

The most frequently asked questions about folding knives refer to the most important part of any folding knife – the blade. In an attempt to cover the majority of these often-repeated inquiries, a substantial volume of text has been included to provide a more comprehensive understanding for the reader, of the folding knife blade with regards to shape, material, the different types and grades of steel (comparative analysis down to the molecular level), points, edges, grinds, finishes (corrosion resistance, etc.) the advantages and disadvantages of each of the above and anything else you ever wanted to know about a folding knife blade but were afraid to ask in public.

Although the blade may be the central focus of any folding knife, sufficient detail in this scope of study has also been provided covering other parts of the folding knife such as the handles, carry clips, opening mechanisms, fastening devices, friction radius, locking mechanisms, lanyard holes and other related hardware.

In addition to detailed coverage of the parts of a folding knife, an equal amount of detailed operational information regarding maintenance, carry, accessibility, opening, closing, locking, unlocking, gripping, rapid deployment, body position, working platforms, application of varying edges and points as applied to flexible and rigid materials, optimal balance, conventional and unconventional working positions as well as other critical components of safe handling and usage of the folding knife are also included.

Folding knives, as problem solving tools, are a pretty darn important piece of utility gear for the first responder who may be called upon to handle situations like suicide prevention, an auto accident, a natural disaster, fire, working on the range, etc., maybe even personal use off duty such as outdoor camping, hiking, fishing or other related activities.

Let's face it we use knives almost every day to open letters, boxes, packages, etc., and depending upon your job responsibilities (electrician, plumber, construction worker, etc.,) maybe used often throughout the day. Uses of the folding knife can include, but are not limited to: normal everyday run-of-the-mill box and letter-opening, cutting twine or tape, outdoor activities such as backpacking, climbing, hiking, camping, skiing, diving, hunting, fishing and of course "tactical" – such as military, law enforcement, federal agency, emergency response (all of which are considered first responders), to name a few.

Scope of Study

Although written predominantly with federal, state and municipal law enforcement professionals as well as US military personnel in mind, this material is applicable to any informed law-abiding citizen (alpha male or not) that has decided to purchase and utilize a folding knife and along with that ownership accept the responsibility for its safe and functional usage.

Model Policy

As a full-time contract instructor for the US military, various US federal agencies as well as state, county and municipal law enforcement, the requests for more detailed information on the usage of folding knives throughout my tenure in the professional training community prompted a tremendous amount of research. The vast majority of the professional training community, for the most part, carries knives. Most of them are folding knives and almost all of them were NOT issued by an employer (except in some very rare cases). Many issues surrounding the selection, carry and usage of folding knives would swirl around conference rooms and bounce up and down the chain of command like a yo-yo as knives were (and still are today for the most part) considered taboo by most upper management.

In particular, such issues referencing usage of the folding knife as a tool versus the folding knife as a weapon prompted serious debate and the handling of the issue as if it were the proverbial hot potato – nobody wanted to touch it not even a little bit, not even today.

As Chief Executive Officer for the Operational Skills Group, LLC (OSG) from 1996 through 2007, it was my responsibility to ensure appropriate training matching federal, state and municipal performance requirements. Serving the Professional Training Community since 1994, OSG offers vetted programs of instruction specific to military, federal and civilian law enforcement requirements. Source-reference ROE/UOF compliant, OSG exceeds industry standards as set by federal regulatory bodies. Such compliances include instructor/student ratios, safety and delivery protocol, outlines/hourly distributions, sustainment, and supporting documentation. (see http://www.opskillsgroup.com for more details)

As a published author and authority on the topic of knives, I was asked in 2000 by Law and Order Magazine to perform a survey of more then 60 US law enforcement agencies nation wide (federal, state and municipal) which was completed in 18 months. All this valuable information was then formed into a proposal which was further developed (with the aid of legal and technical assistance) into a Model Policy, Concepts and Issues Paper and Training Key (#576) and submitted to the IACP for final approval.

On December 2004, the IACP Commissioners accepted and approved the policy and recommended that I write the official Training Key #576 which was then added to the IACP library. As of this writing, the IACP Model Policy for the carry, deployment and safe usage of knives by peace officers including the IACP Concepts and Issues Paper (supporting documentation) and IACP Training Key #576 are now available through IACP (www.iacp.org).

While conducting research to develop this new policy, I worked closely with Buck Knives, Inc. and Strider Knives, Inc. to develop products that would meet the needs and requests of law enforcement officers. This collaboration led to the development of a series of police knives, first released in August 2003. Following years of continued research, development, testing and most importantly feedback from the field, I have continued to work for the benefit of the US military and law enforcement community with similar folding knife development projects.

What exactly is IACP (International Association of Chiefs of Police)? Founded in 1893, the association's goals are to advance the science and art of police services; to develop and disseminate improved administrative, technical and operational practices and promote their use in police work; to foster police cooperation and the exchange of information and experience among police administrators throughout the world; to bring about recruitment and training in the police profession of qualified persons; and to encourage adherence of all police officers to high professional standards of performance and conduct.

Since 1893, the International Association of Chiefs of Police has been serving the needs of the law enforcement community. Throughout those past 100-plus years, they have launched historically acclaimed programs, conducted ground-breaking research and provided exemplary programs and services to their membership across the globe. They are considered the authoritative voice to the nation's law enforcement agencies.

Professionally recognized programs such as the FBI Identification Division and the Uniform Crime Records system can trace their origins back to the IACP. In fact, the IACP has been instrumental in forwarding breakthrough technologies and philosophies from the early years of their establishment to now, as we move into the 21st century. From spearheading national use of fingerprint identification to partnering in a consortium on community policing to gathering top experts in criminal justice, the government and education for summits on violence, homicide, and youth violence, IACP has realized its responsibility to positively affect the goals of law enforcement.

What are IACP Policies? In 1987, the International Association of Chiefs of Police entered into a cooperative agreement with the U.S. Justice Department's Bureau of Justice Assistance to establish a National Law Enforcement Policy Center. The objective of the center was to assist law enforcement agencies across the country in the critical and difficult task of developing and refining law enforcement policy.

Organized under the direction of a broad-based advisory board of recognized law enforcement professionals, the center has carried out its mission through the development of a wide variety of model law enforcement policies. Each model incorporates the research findings, the input of leading subject experts and the professional judgment of advisory board members who have combined this information with their extensive practical field and management experience. The end product is some of the best contemporary thinking in the field.

The policies addressed by the center are selected because they represent some of the most difficult issues facing law enforcement administrators. The policy center continues to develop models in other priority areas.

Folding Knives: Carry & Deployment

What are IACP Training Keys? Concise, authoritative sources of law enforcement information, these six-page, loose-leaf monographs allow law enforcement officers to expand or sharpen their knowledge, skills and abilities on a broad variety of law enforcement practices and procedures. Each Training Key is prepared by a leading expert and addresses an issue of particular interest to line officers and their supervisors.

For well over 35 years, local and state law enforcement agencies have looked to the Training Keys for the most current information in the science and practice of policing. Ideal for roll-call training and formal classroom instruction as well as independent study, each one includes questions and answers to test and document student learning. All sworn officers can expand their professional law enforcement knowledge by using the Training Keys systematically and routinely.

The Training Keys have been priced so all law enforcement agencies can afford them: For a small annual fee per officer, each officer receives 12 individual Training Keys enough material for many departments to meet a full year of in-service training needs. Also available are over 350 additional Training Keys, published in the past 25 years, covering over 400 topics, including accident, criminal and death investigation, arrest procedures, arson and bombs, burglary, interrogation and interviews, juvenile delinquency, legal procedures, narcotics and drugs, special tactics, traffic enforcement and control, and vehicle theft, among others.

Concepts and Issues

The Knife Safety and Usage Concepts and Issues Paper (which was required as support documentation for the knife policy) was prepared and submitted by yours truly in December 2004 to accompany the Model Policy on Knife Safety and Usage published by the IACP National Law Enforcement Policy Center. The paper presents essential background material and supporting documentation to provide greater understanding of the developmental philosophy and implementation requirements of the model policy. This material proved valuable to law enforcement executives in their efforts to tailor the model to the requirements and circumstances of their community and law enforcement agency. The following are a few excerpts from that original paper.

Carried as part of an officer's uniform, in plain view of public and used by the officer in public, a majority of police officers carry knives both on and off duty. Interesting enough, a general Police Utility Knife safety and usage policy does not exist for a vast majority of departments. Conversely, such general policy, exists for carry, deployment and usage of other equipment either issued or available to the officer, such as handcuffs, baton, irritant propellants, firearms, communication equipment, etc. However, without such policy, an officer may carry in plain view of public and operate said equipment (specifically the knife) within the guidelines of his / her own discretion. This places the onus of selection, safety and usage on the officer and consequentially on the department as the employer and trainer of the officer.

The very topic of knives is considered taboo to most administrators. Regardless, the stark reality is that a predominant number of officers currently carry knives. In fact, police officers carry a variety of different knives and carry them in a variety of different ways. Without policy, there are no prohibitions or limitations other than surface area on the belt or uniform or supervisors who may be particularly mindful of public view.

As a practical matter and by convention it is an accepted fact that a police utility knife is a very necessary piece of equipment in police work. However, with no written policy or statute for support, most departments fall back upon or default to state guidelines referencing equipment and application – none of which address the police utility knife or its usage. Insufficient training, minimal experience levels and lack of standards or guidelines lend any agency wide open to various degrees of negligence that will eventually lead to more serious consequences. A sound policy in place providing guidelines of governing both the quality and handling of the Police Utility Knife, clearly demonstrates such standardization and guidelines with training provisions and would in fact prove due diligence and certainly curtail the potential of such problematic issues as negligence.

Certain administrators argue that the onus of responsibility rests squarely upon the shoulders of the officer as the administration does not want to incur the responsibility of providing written policy. Such arguments are swiftly rebutted when an incident becomes public and administration is held liable for the negligent action(s) of their employees which occur within the scope of their employment as police officers. For example, if an officer uses a low-quality folding knife to pry open a window and the knife breaks and injures an innocent bystander.

A sound department policy supports the agency in any post-incident review, and demonstrates that the department recognizes the need for officers to be properly equipped and trained to handle the wide variety of responsibilities they may encounter.

Intended Usage: The usage of any Police Utility Knife is for utilitarian purposes only. Any references to the knife as a "weapon of self-defense" or "last-ditch survival tool" are not recommended. Any questions regarding the use of a knife as a weapon should be referred to standard department Use-of-Force policy. The police utility knife, like an officer's handcuffs, radio, flashlight or ballpoint pen, can be used as a weapon of opportunity but first and foremost should be intended for use as a tool. This policy is designed to reflect the role of a knife as a utility tool and not as a weapon.

Quality: Carrying and using a substandard or low-quality utility knife could cause potential problems for the officer (possibly cutting himself or another) and the department (accidentally cutting a suspect as in the case of Detroit Police Dept. April 2003). On the other hand, there's no need to either issue or authorize costly or overly expensive knives.

In one reported incident, a hanging situation (Bloomington, Indiana, 2002), it was a race against the clock to try and save the person. Officers flooded into the room frantically reaching for knives. Unfortunately most of the first responders did not have a knife on their person and they had to wait for the ones that had a knife on their person to cut the individual down. The need for first responders to carry a knife is plainly obvious and should be written into policy.

Carry and Retention: While some states may allow a concealed knife under the vest, or a boot knife, generally most departments discourage carry and deployment of a double-edged dagger. Retention of the knife while interacting with the public is also a concern. In more than one reported incident, a knife fell out of its sheath in the middle of a non-compliant arrest, which resulted in a shooting. Poor equipment, poor training and poor carry technique resulted in the subject arming himself with the officer's knife.

Appearance and Quality: Regarding size and appearance, the problem with carrying a large Bowie Knife conspicuously on your gun belt is that it wouldn't be well received by the public. No exact length or overall specifications have been established nationwide; however, a general policy referencing product quality, maximum blade length, safety and training would set standards and provide such guidelines for appearance and quality. Specifications provided as part of the Model Policy are the result of nation-wide polling of officers of all ranks and responsibilities within the federal, state and municipal law enforcement community (interviews and articles by the author for Law and Order Magazine, Jan 2002).

There are numerous models and makes, however, there are certain considerations which cannot be overlooked such as quality, size and functionality. For example a "fighting knife" generally has a thinner blade and cannot be used effectively for utility purposes. The purpose of these specifications is to provide some guidelines as to ensure integrity of the blade in quality and functionality.

Many officers (more specifically tactical team members) are concerned with accessibility and retention of their fixed blade and the durability of the sheath. All of these are excellent points and should be addressed in policy. In one incident, during a high-speed vehicle pursuit, the officer driving had a boot knife in a sheath in his boot. Not securely fastened and carried in a substandard sheath, the knife fell out during the rigors of the pursuit. Unbeknownst to the officer, the handle landed on angle against the floorboard with the tip pointing upward. The next time he slammed his foot on the brakes he inadvertently drove the blade more than an inch and a half into the calf of his right leg.

Any officer is trained in the carry, deployment and usage of all other tools on his/her belt and uniform, such as handcuffs, baton, irritant propellants, firearms and communication equipment, but not the utility knife. Conversely, there is little or no training available to these officers in the carry, safe handling and proper usage of the police utility knife – a common tool used in police work. Such a training course should consist of not less than four (4) hours of initial instruction and periodic updates, for example, during in-service training courses or periods of requalification.

Since, the very topic of knives is considered taboo by most administrators (regardless of the hard fact that most officers carry a knife and that it is part of his/her uniformed presence, in plain public view and is considered one of several tools available to that officer for police work), training in the carry, safe handling and usage of a knife is often quietly overlooked. It is therefore important to note that insufficient training, minimal experience levels and lack of standards leave an agency wide open to varying degrees of negligence and consequential liability. As unlikely as a Civil Rights (Section 1983) lawsuit liability may be (City of Canton v. Harris, 489 U.S. 378, 109 S. Ct. 1197 (1989) - briefly summarized, a department may be culpable for sending their employees out into the public untrained) by alleging failure to train are – having a policy in place will certainly prevent those lawsuits from being viable. Another important point is that appropriate training will also reduce the number of worker compensation claims.

General policy regarding product quality, maximum blade length, safety and training is strongly recommended. Any type of policy that would fit into the general use-of-force policy in that broad spectrum of application, would work for general purposes.

Part I
Cro-Magnons to Cryogenics

Part I

Folding Knives - A Brief History

Knives have been around since before the beginning of recorded history. Various mainstream archeological digs have unearthed evidence of man's usage of knives dating back plus or minus 8,000BC and some research even prior. The historical evidence throughout the evolution of mankind with regards to development of early edged tools demonstrates that the knife was one of the very first hand-held tools ever created. No one really knows if the knife's first use was as a weapon or as a tool. In fact, more than ten thousand years later the debate still rages on today with regards to the development of modern knife usage (weapon or tool) policy for US law enforcement agencies.

As with all useful inventions, the knife was a pretty good idea and withstood the test of time – lots of time, roughly ten to possibly twenty thousand years worth of time. Overall knife design went through various stages of development. The very first materials were most likely a sharpened bone or stick. Later on chipped stone and eventually modern metallurgy came into play.

The very first knives were of course "fixed" – meaning that there were no moving parts. An early knife user simply reached for his fixed knife and pulled it out (hopefully by the handle) and began using it. Throughout history, the knife proved to be such an effective and efficient tool that it was further refined and developed. New variations were added such as a more comfortable handle and more sturdy materials which were able to hold an edge longer. Later on down the timeline sheaths were

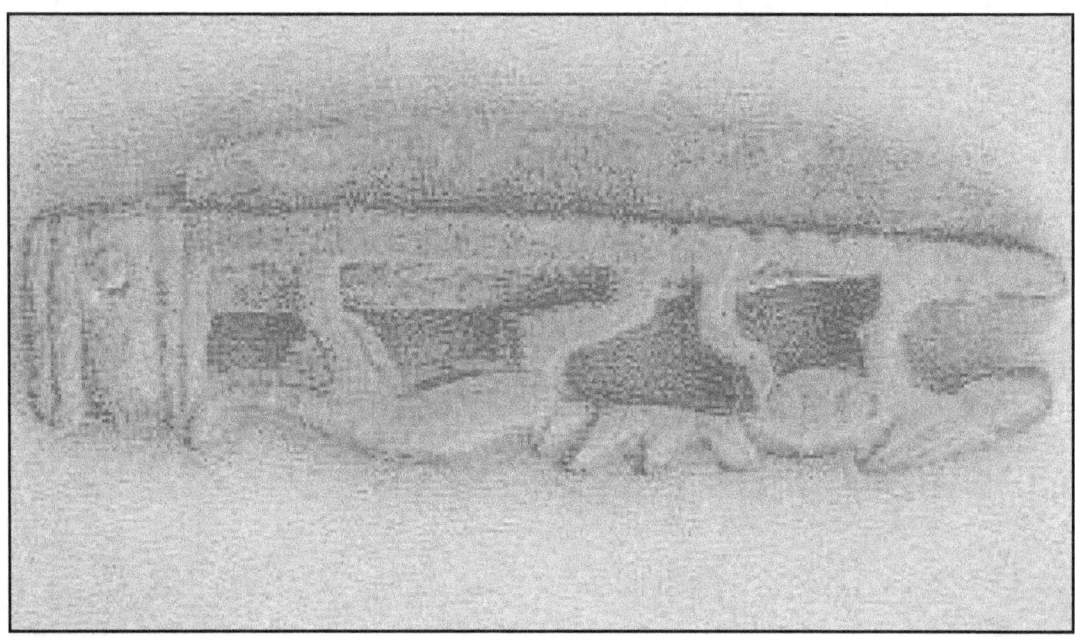

Ancient Roman Pocket Knife

invented so that the knife could be carried by the owner for both convenience and in such a manner as to not inflict any unintended bodily damage upon himself or his neighbors.

By the time of the Roman Empire knives were very well-developed and in addition to functionality, aesthetics had entered into the development equation. New innovations flourished in the manufacture of knives until roughly about the 2nd or 4th century AD when, according to archeological records, folding knives where introduced.

Although fragments were found sporadically from various cultures, the first complete folding knife unearthed demonstrated clearly advanced workmanship. Drawing upon both functionality as well as aesthetics, these early Roman knife-makers developed one of the world's first truly functional folding knifes.

Elsewhere on the planet there were other innovative ideas, such as carrying the blade between two broken pieces of bone. The concept of "knife between broken bones" originated in the Philippines. The term "Bali" (like the island of Bali located due east of the island of Java in the Indonesian Archipelago) literally translates to the

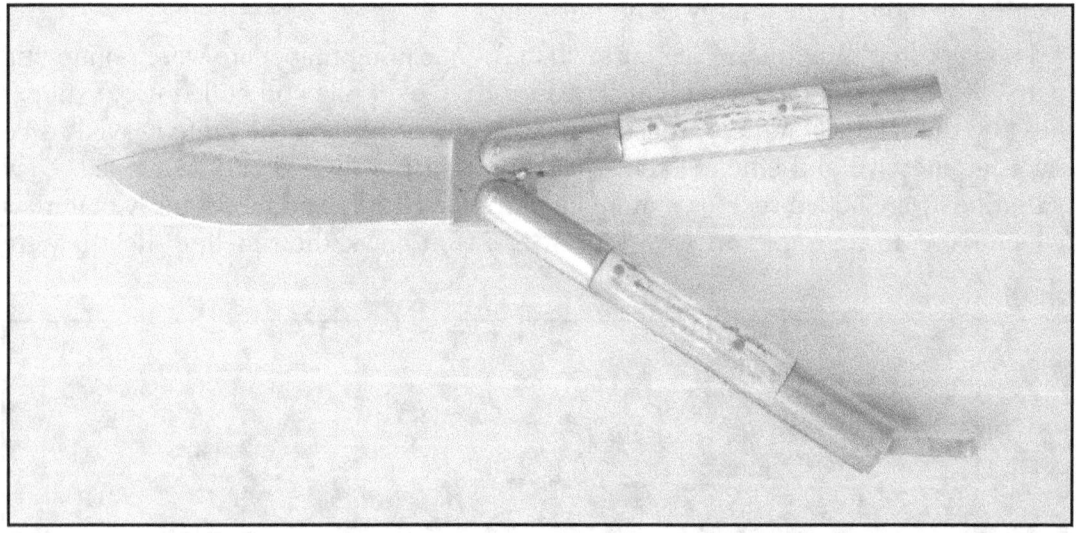

Knife inside "broken bone" – Balisong

word "broken." The term "soung" is from a dialect which means "bone" – the two terms used together literally translate to "broken bone" or more appropriate when referring to the folding knives of the Balisong Barrio in the Philippines "knife hidden in broken bone.'

Categories of Knives

Today, in the knife world there are tens of thousands of models, styles, blades and handle configurations. At the one end of the spectrum there are big long knives that border on "short sword" and at the other end of the spectrum there are extremely small customized hand, finger and toe knives which border on the edge of "improvised weapons." The world of knives incorporates any and all in between.

The world of knife-making is additionally divided into two general partitions of manufacturing: custom knives and production knives. The difference between the two is that custom knives are for the most part hand-made, that is manufacturing of the knife without the use of modern "tooling" and mass-production processing. As a result of this time-intensive and hands-on process, the cost to make a custom knife is higher.

Production knives, on the other hand, are mass-produced and involve the use of factory tooling and assembly processes which reduces overall costs and in the end allow the buyer more affordability.

Regardless of manufacturing considerations and in order to further classify the immense volume of such a wide and diverse set of cutlery, it becomes necessary to first break down this massive hodgepodge of varying sizes, types, materials, styles, etc., into two general categories of knives.

Fixed and Folding

In the knife world there are two general classifications of knives – these are commonly referred to as "fixed" and "folding." The term "fixed blade" is applied to the type of knife that has no moving parts. A "folding blade" is that type of blade which is defined by the fact that it has moving parts. "Fixed Blade" is synonymous with the term "Fixed Knife" as "Folding Blade" is synonymous with the term "Folding Knife" – there are advantages and disadvantages to both categories.

As with all edged tools there's a plus and a minus corresponding to every feature. It has been said that "good or bad - everything has a price tag" this axiom can also be applied to the world of knives, specifically the difference between fixed and folding blades.

Referencing a fixed blade, the plus is that there are no moving parts on a fixed blade which by default makes it more reliable than a folding knife, but the price tag is that in order to carry a fixed blade you need a sheath and it requires that you wear that solid piece of steel somehow affixed to your body when sitting in your car, driving around on patrol, going to meetings, etc.

The folding knife is generally much smaller and therefore more convenient to carry than its fixed counterpart. However the trade-off is that the blade is small and it does in fact have moving parts which render it by default less reliable than a fixed blade.

Fixed Blades

A fixed blade can be defined as any pointed or sharpened, single or double-edged blade secured to a fixed handle. Well-known examples of a fixed blade would be the classic Bowie Knife, Scottish Dirk, American K-Bar and the classic British "Commando" double-edged dagger. Even a broken piece of glass or steel shank with duct tape wrapped around one end would classify as a fixed blade.

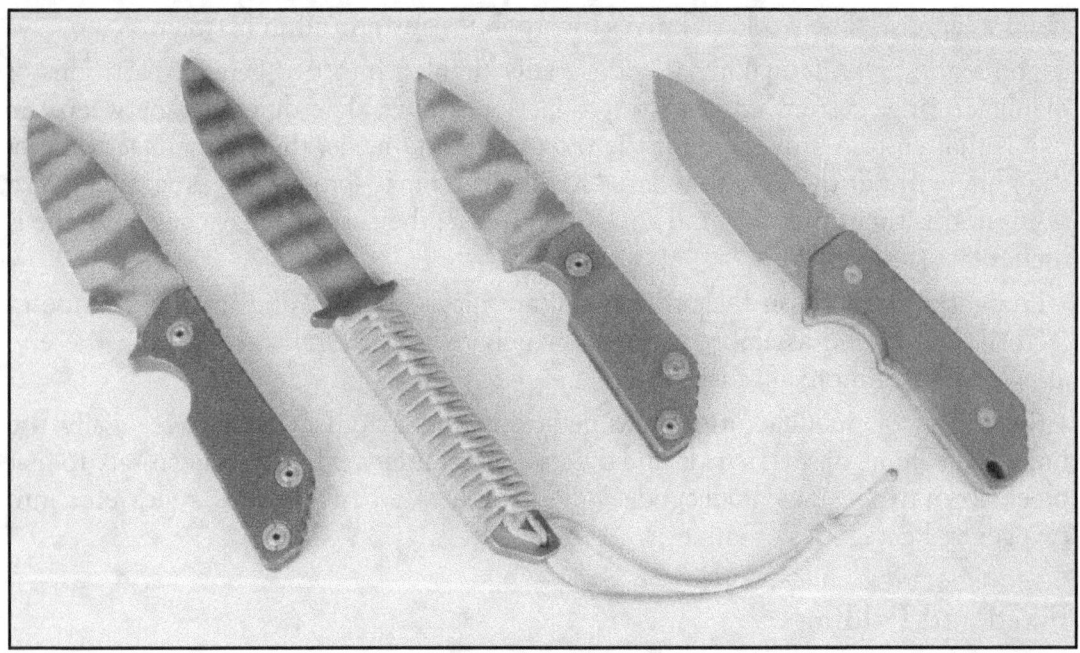

Examples of high quality fixed blades – courtesy of Strider Knives Inc.

Although a great knife to have with you in the event of digging a hide (if you are a sniper), tearing into drywall (if you're a professional law enforcement specialty team member), prying into a window or door (if you are forced to make an ad hoc entry), etc. one must keep in mind public appearance. How would a fixed blade be perceived by the public – especially if you were a patrol officer with a major metropolitan police department? What about going to a meeting with a Bowie knife tucked in behind the waistband of your suit pants? These are important considerations when deciding between fixed and folding.

Folding Blades

A folding blade can be defined as any pointed or sharp, single or double-edged blade which in any way can be folded, coiled, bent or otherwise secured in such a fashion as to be rendered disabled or "unlocked" in the "folded" position. Examples of a folding blade would be a pocket-knife such as a Swiss Army, Boy Scout (or even Cub Scout) knife or a "combat folder." Switchblades and "combat" automatics also qualify as folding blades.

Part I

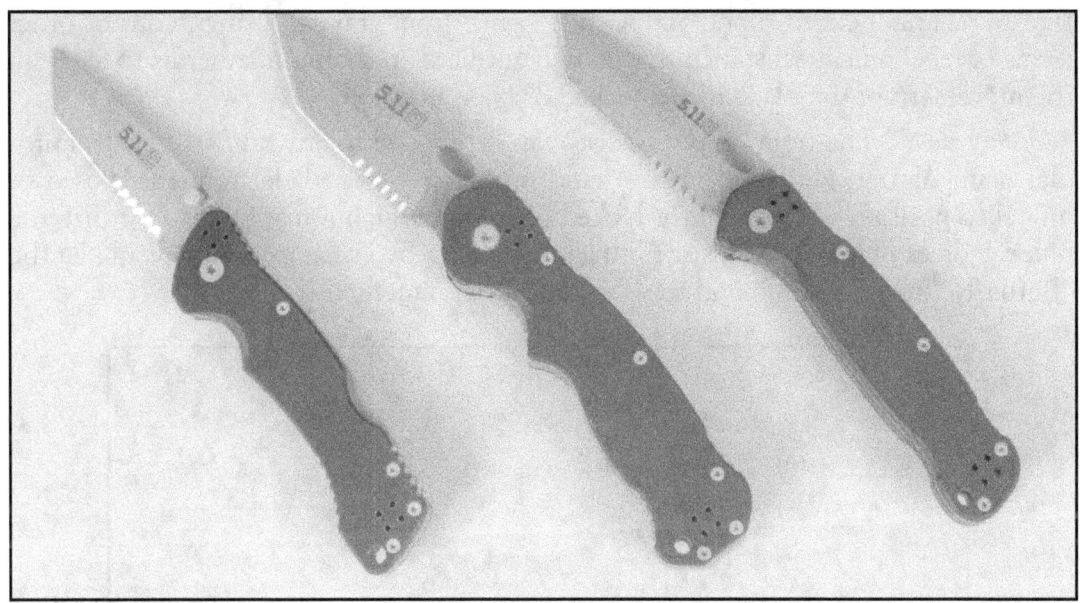

Examples of high quality production folding blades – courtesy of 5.11 Tactical.

Mechanical

In the world of folding knives there are only two mechanical categories – "assisted" and "unassisted."

Assisted. The term "assisted" applies to any folding knife where anything other than the usage of the human hand "assists" in the moving of the blade from the closed or unlocked position to the open or locked position. Some examples of this assistance would be say "spring assist" and "gravity assist."

Unassisted. The term "unassisted" generally applies to those folding knives which can ONLY be moved from the closed or unlocked position to the open or locked posi-

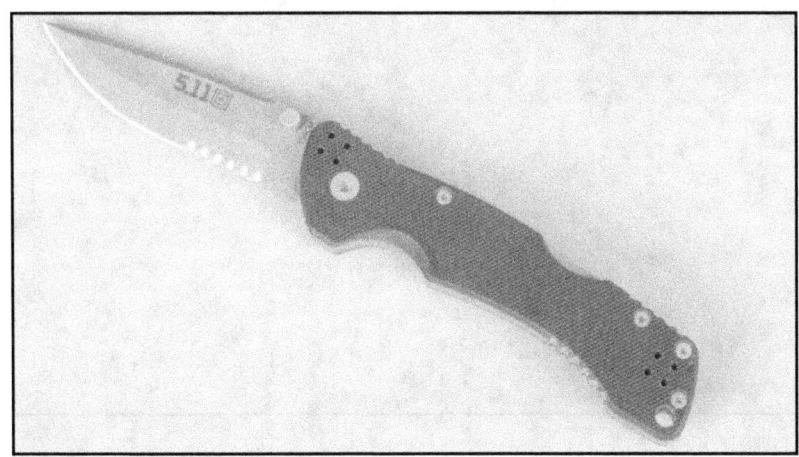

Example of a mechanically "unassisted" folding knife.

tion by "manual effort" alone and with no assist from gravity, springs, coils or other forms of mechanical assistance that may be utilized in such a manner as to assist in the movement of the blade into a locked or open position.

Gravity Assist. The term "gravity assist" can be applied to any knife that with minimal manipulation with the human hand, may move the blade from the closed or unlocked position to the open or locked position in such a manner as to utilize the natural forces of gravity to assist in this movement. Two such examples would be the "Butterfly" (aka Balisong) and the Tri-fold models among others.

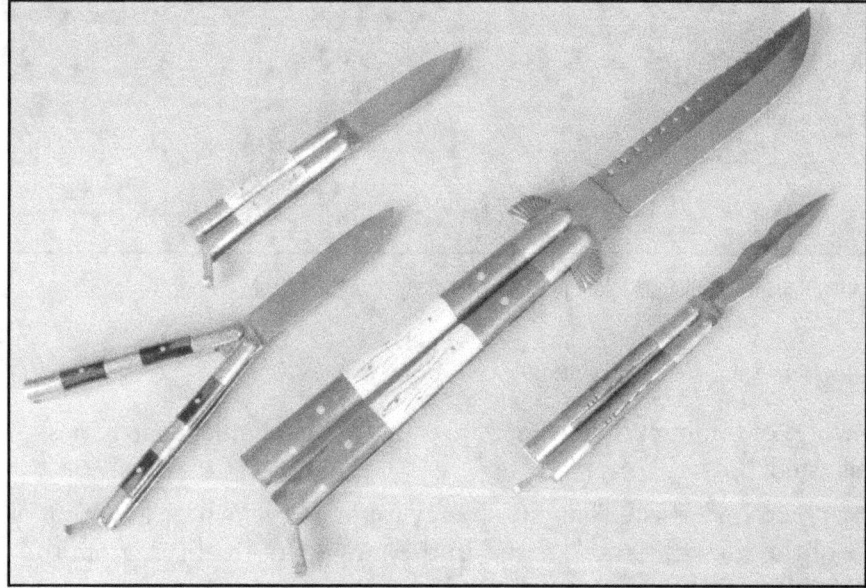

"Butterfly" or "Balisong" folding knives.

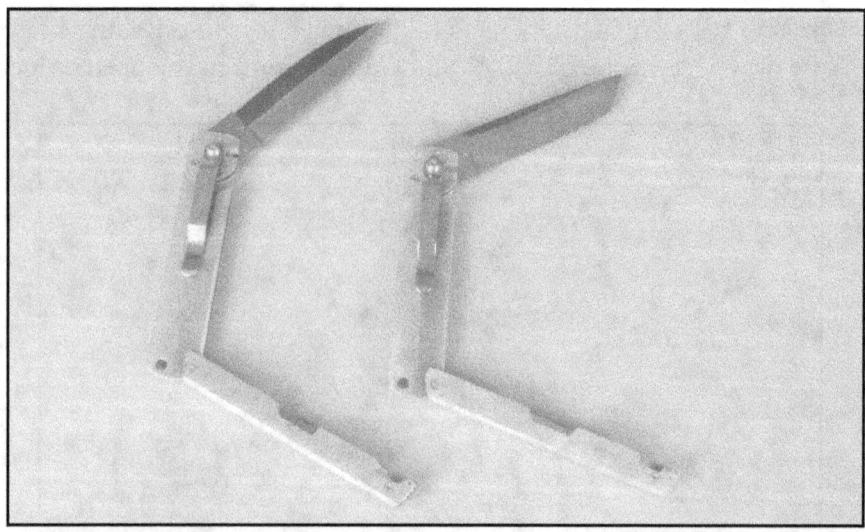

Examples of Tri-folds

Spring Assist

The term "spring assist" applies to those folding knives where the application of a spring or springs in either coiled or uncoiled configuration are utilized in such a manner as to in any way assist or partially assist with the movement of the blade from the closed or unlocked position to the open or locked position. In the world of spring-assisted folding knives there are generally two categories of classification: 1. Automatics (aka "autos") and 2. Semi-automatics.

Autos. The term "Auto" is an abbreviation of the word "Automatic" which is applied to any folding knife which utilizes some form of spring system by which to move the blade from the closed or unlocked position, all the way to and including the open or locked position. A classical example of an auto would be a switchblade.

Semi-autos. The term "Semi-auto" is an abbreviation of the term "Semi Automatic" which is applied to any folding knife which utilizes some form of spring system by which to partially move the blade from the closed or unlocked position, part of the way toward an open or locked position. A classical example of an auto would be a "spring-assist semi-auto." Generally these types of knives have additional moving parts in or as part of the handle (sometimes hardly noticeable) which assist in the release of the spring in such a manner as to assist with the movement of the blade.

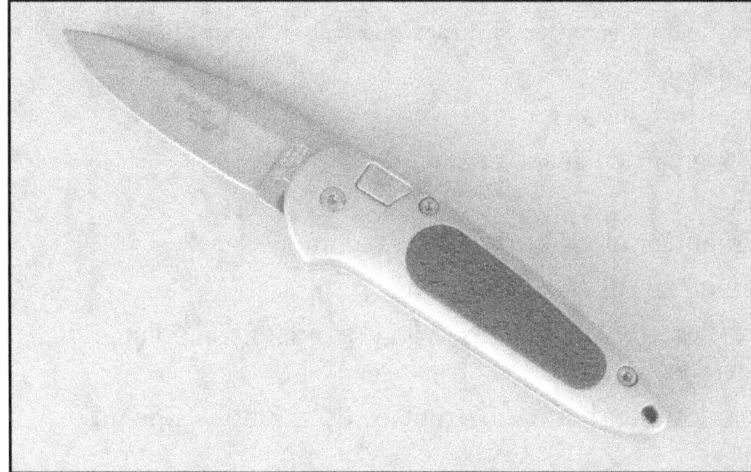

Example of an "Automatic" spring-assisted folding knife

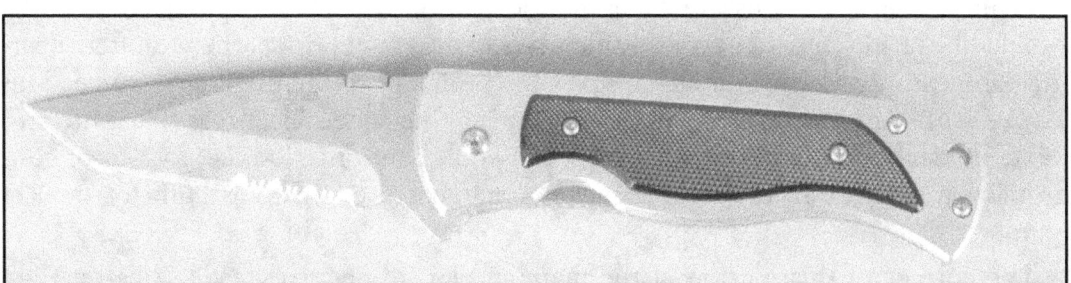

Example of a Semi-automatic mechanical folding knife.

Metallurgy and Blade Geometry

Contributed by Duane Dwyer, Co-Founder of Strider Knives Inc.

Before we get into the meat and potatoes of Metallurgy and Blade Geometry, let's first take a closer look at the words "Technical and Tactical" as they pertain to this text. According to Encarta® World English Dictionary (© 1999 Microsoft Corporation) the definition of the two words are as follows:

Tactical (tac·ti·cal) adj

1. relating to or involving tactics
2. done or made for the purpose of trying to achieve an immediate or short-term aim
3. showing skillful planning in order to accomplish something
4. used or made to support limited military operations
5. undertaken or for use in support of other military and naval operations

Technical (tech·ni·cal) adj

1. relating to or specializing in industrial techniques or subjects or applied science
2. skilled in practical or scientific subjects
3. belonging to or involving a particular subject, field, or profession
4. exhibiting or deriving from technique or the use of technique
5. according to a strict interpretation of rules or words
6. used to describe a type of security analysis based on past prices and volume levels as well as other market indicators
7. used to describe outdoor clothing that has been made using state-of-the-art materials and techniques

While these definitions are accurate and descriptive, they are also broad and I would like to add that for every specific usage of a term or phrase, there is often a particular meaning, which is usually a fine tuned element of the original resource. This is true with these two words with the following fine tuned meanings related to this text: Tactical: Skills, mindset and/or tools used in confrontation, i.e. for war or confrontation. Technical: measurable sciences applied to edged tools built for tactical purposes.

The purpose of this portion of the manuscript is to give the reader a better than average understanding of the technical aspects involved in our subject matter. At the end of this chapter the reader will either have a greater margin of success at choosing an edged tool appropriate for their need or will be sound asleep. To prepare for the latter, I recommend a padded toilet seat to reduce the numbness in the legs after awakening.

The reader will notice various source references throughout this chapter. It is my hope that the reader would please understand that even if the specific name of a contributor is not mentioned, a well-deserved and sincere "thank you" goes out to all of those who have helped evolve our base of knowledge in this technical field over the years.

Our former Commander-in-Chief, the Honorable Woodrow Wilson, once said; "I not only use all the brains that I have, but all that I can borrow." I believe the President was referring to the use of subject matter derived from subject matter experts.

First, what is a knife? Simply stated by United States Marine Corps Colonel Robert Coates, "the knife is a problem solving tool." As both a child and an adult, I have often violated the age-old rule of "get the right tool for the job" for sake of saving time. In a vast majority of cases, we find that not only would we have saved time with the "right tool" but we would have also done a better job in the end. Therefore, the key is in identifying the problem.

The first step to solving any problem is to identify the situational needs, thereby distinguishing the critical requirements for a solution. Once we have established what we need to accomplish, we can apply known science and equip ourselves with the greatest margin of success at accomplishing the intended task. In many cases, we find ourselves limited to a specific tool and confronted with new tasks where the lengthy issues of design and manufacture are not an option. We simply have what we have. This is where good fundamental choices will pay off and the information to follow will help with those choices.

Now that we have identified our needs, our goal is to analyze the aforementioned tasks involved in the particular mission statement/environment where the edged tool will be utilized and design a configuration that will equip the user with the greatest margin of success for that mission statement/environment. Our next step is to isolate the truly technical aspects that are applicable to our tool. By "truly technical" I am referring to issues that science has determined test methods that utilize, measure, record and compare data in a standardized fashion that when tested by multiple parties will achieve the same results. This gives the world great foresight in designing stages. It produces results, which are not arguable.

Beware of "back yard" tests boasted by companies and individuals. These types of tests are valuable in some cases, however, they produce LITTLE or NO legitimate data for comparative analysis. Another former Commander-in-Chief, the Honorable John Quincy Adams once said; "Facts are stubborn things; and whatever may be our wishes, our inclinations, or the dictates of our passions, they cannot alter the state of facts and evidence." Touchdown for science!

There are three primary technical variables involved here. We will address each of the three in the following order:

1. Raw Material Selection
2. Exploitation of Raw Materials
3. Geometry

First, let's address the raw material selection. Common choices in this area include but are not limited to:

1. Carbon based alloys
2. Stainless alloys
3. Non-ferrous alloys
4. Non-conductive materials

Each of these has its advantages and disadvantages and while a thorough description of each would be far too lengthy for this text, examples of each are suitable and helpful here.

From the carbon based alloys we have centuries of edged tool making with a range of simple materials easily found even in lesser developed areas of the world such as leaf springs from vehicles and old hand files up to and including selections within the highly sought after particle metallurgy grades. There are many excellent choices here for edged tools across the board with the primary criticism being the corrosive nature of this class of materials.

From the stainless alloy class, it was shortly after 1900 that English metallurgist Harry Brearley is credited with having developed stainless steel in his search for an alloy to protect cannon bores from erosion. The first commercial production of stainless steel occured in August 1913. There were many other individuals working on chromium alloys around the same time period but generally speaking, credit is given to Mr. Brearley.

Another milestone in 1963, stainless alloys were first applied to razor blades and edged tools had a new face. (Ref.: Specialty Steel Industry of North America SSA Association)

Like the carbon based, non-stainless alloys, there is a wide range of materials in this category as well, ranging from simpler, dated alloys to the exotic materials produced utilizing the above mentioned particle metallurgy. This class of material took a justifiable amount of criticism in its infancy (within the edged tools world) due to its limited capabilities and difficulty to be ground and polished. Quite contrary to this now are the magnificent, very complex super alloys produced and going strong in the world of edged tools.

The "particle metallurgy" referenced above more specifically defined is one of the greatest single leaps made by technology since the Iron Age. It is properly referred to as the CPM Process or Crucible Particle Metallurgy Process. This technique is a complex combination of several patented applications. It was first developed in 1970 and is the current standard when high alloy grades desiring improved wear resistance, toughness, grindability, heat treat response, thermal stability, stable substrate for coatings, and efficient wire EDM cutting are sought after. The process itself could fill chapters if discussed in depth so we will hit the high points.

The CPM Process takes homogenous molten bath, and bursts the material through an atomizing nozzle where the spray forms into tiny spherical droplets. These solidify and collect in the bottom of an atomization tower, relatively spherical in shape and uniform in composition. Each particle being a micro-ingot, solidified so rapidly that segregation has been suppressed. (This results in extremely fine carbides which endure throughout the mill processing and remain fine in the finished bar.) The powder is then loaded into steel containers and vacuum-sealed. These containers are then hot isostatically pressed (HIP) utilizing pressure and temperature. The resultant microstructure is homogeneous and fine grained with uniform distribution of the tiny carbides thereby alleviating alloy segregation and preventing grain growth. (The majority of data provided on CPM Process can be found in reference manual titled "Crucible Tool Steel and Specialty Alloy Selector").

Third on our list are Non-Ferrous alloys. These are a substantial minority within the edged tool realm, but need to be included here because of a few key issues. Non-ferrous alloys include a vast multitude of materials, however, very few find use as edged tools. Examples are non-sparking alloys for work that is sensitive to electrical discharge. Lightweight alloys, primarily Titanium, have been used in the edged tool world with limited success due to the lack of the ability of these materials to obtain comparable technical characteristics to carbon and stainless steel alloys. Having a great deal of interest in this Element in its alloyed configurations and what it potentially has to offer, we spent substantial time researching ideas in hopes of eventually seeing a Titanium alloy fit the edged tool category.

We were fortunate enough to work with the experts at the Crucible Materials Corporation on a new material in this category from its genesis. The staff at Crucible is no strangers to the world of Titanium. Alloy 6Al4V, arguably the most important titanium alloy was patented by Crucible in 1956. It was then and has remained through today one of the most commonly used of all the titanium alloys in aircraft and prostheses usage.

Crucible's development of a new material has now achieved a level of success great enough to acquire a patent from the US Patent Office and reach hardness levels comparable to specialty steel alloys (above 60 R/c). What this means for the user is the realization and availability of a material lighter, non-magnetic, equally as hard and completely corrosion free in the near future. This material is still undergoing its initial phase of rapid evolution and more information will hopefully soon be available. The really great news is that after approximately 4,500 years of the Iron Age, this could be our ticket out!

Having had the opportunity to work with this material in the development stage is yet another reason for my personal accolades of the loyal, professional and responsive group at what we like to call "Team Crucible."

For those who are history buffs we have provided the following bit of very interesting information on this organization:

Crucible Materials Corporation was created in December 1985 following an employee buyout from Colt Industries. This was a return to independence that started in 1900 with the incorporation of 13 steel companies into the Crucible Steel Company of America. The result was a strong quality oriented specialty steel company.

Crucible's history dates back to 1876 when Sanderson Brothers of Sheffield, England established a plant in Syracuse, New York. Sanderson became one of the 13 plants forming the new corporation. It was at this location in 1906 that steel from the first electric arc furnace in the United States was produced.

Crucible Research was founded in 1929 and remains the center of Crucible's development of new processes and products, improvement in the quality of existing products, and the solution of production and consumer problems.

The company produced the world's first commercial powder metal tool steel in 1971 at the Syracuse plant. The powder technology was expanded to include superalloy powder production in a facility built in Oakdale, Pennsylvania in 1976. A titanium atomizer was developed in 1984 making Crucible one of the most technically advanced producers of powder in the world. The reason for the abnormally large recognition of an individual nature is twofold; first, it is deserved scientifically, and second, this is the first actual steel producing organization and accompanying mill ever to work hand in hand, case after case with the edged tool industry both on the giant production company level as well as on the small specialty company level without hesitation. (The data provided on Crucible Materials Corporation can be found at their website: www.cruciblematerials.com).

Now back to our fourth and final choice in our "Raw Material" discussion.

Non-metallic materials: These have also been included here as they are utilized in the edged tools category, primarily where conductivity is a concern. Examples of these materials are flint, carbon fiber, glass, glass epoxy substrates, ceramic, wood, micarta using phenolic resins, various grades of plastic and nylon as well as many others. Their specialty attributes vary from non-conductive needs to historical applications and more. These materials are, though a less technical facet of our discussion, still a recognized constituent within the subject matter.

This concludes the brief look into our first of three primary technical variables, raw materials, and brings us to the second variable, exploitation of raw materials.

Some have asked why the term "exploitation" is utilized here. Well, other terms that come to mind are limiting where the term exploitation, in this case, means "the use or development of something to produce a benefit." (Encarta) This is a broader and more applicable description for our purpose.

Four million years ago, and in some places today, chipping stone (better described as flint knapping) was and is an example of exploitation. More examples are fire and snow (like we all saw Conan's father do), working with iron ore after the demise and fall of the prior Bronze Age. Also, forging via hand held hammers or modern pneumatic and hydraulic hammers which result in better application through selective applied grain structure, differential tempering and other facets affected through this process. There are examples of this type of exploitation enough to fill volumes, however, to forsake the numbness in the legs again, we will move on!

More specifically, the current application of exploitation lies in the comprehensive and elaborate recipes involved in maximizing the potential of the "super alloys" that we currently recognize as leading edge technology. These materials are manufactured (for the most part) and supplied in an annealed state for ease of manufacturing. Once the desired dimensional configuration is achieved, the exploitation begins. This multi-stage process is reliant upon knowledge, equipment and application. The knowledge is made up of temperatures, times, gases, solids and liquids. The equipment is built, at great expense, to transform materials of this nature to their desired state. Much like water can be manipulated to create ice or steam, the "super alloys" of today can assume many forms. Precise, computer controlled manipulation achieves desired end results in order to achieve the original requirements. Factors that are affected here are the very aspects involved in the original material choice for the application at hand. In short, the exotic materials will not reach their maximum potential without these sophisticated procedures in this evolved, highly technical equipment.

In analyzing the effects of the first and second technical variable combined we find that once again we achieve a short list. This list is comprised of three variables that can be massaged to fine tune the combination of a material choice with a perspective exploitation recipe to fit a requirement.

Here are three properties of these materials, which may be varied independently through exploitation to some extent:

1. Hardness-resistance to deforming and flattening
2. Toughness-resistance to breaking and chipping
3. Wear Resistance-resistance to abrasion and erosion

The importance of including this portion of data is to inform the reader of the intense and complex nature of this vital combination of properties within another combination of technical variables.

Bottom line… countless hours of research and development, destructive testing, nondestructive testing and the compilation and publication of results along with applied changes through evolution, feedback and trial and error are all part of the makings of the next great material.

Folding Knives: Carry & Deployment

Now we'll take a look at our third technical variable: geometry. This one is simply described. One does not shave with a splitting maul, nor does one chop wood with a razor blade, however, the cutting geometry of most every edged tool imaginable lies between these opposite ends of this spectrum. Edge angles and blade grinds (flat grinds, hollow grinds, flat chisel grinds, convex grinds, etc. which will be covered in more detail later) have been and will continue to provide us with a source to one-up each other for decades and possibly centuries to come. In many cases this choice is made due to manufacturing limitations. In the few rare stated cases where an individual or organization specifies a superior method, we will usually find that in due time, a newer, better, more superior combination appears. The real truth is that this technical variable is as great a contributor to our finished product as the first two. Therefore, rather than quibble over what to do on any single aspect of these three variables, we recommend that one work with the most modern basic fundamentals available that are... ready? ... TRULY TECHNICAL, along with everyone's overriding and controlling factor: cost and make an educated, well thought out purchase.

A few additional notes on related considerations: Though neither truly technical, nor comparable data, these are very important aspects regarding edged tools which the reader will see integrated further into our study of folding knives:

1. Carry and Carry Systems (Carry Clips, Sheath/Scabbard/Holster)
   - Sheaths are applicable to both fixed and folding blades.
   - Must meet mission specific needs/requirements
   - Durability
2. Ergonomics
   - Overall size issues compared to abilities of user
   - Usage orientation (This refers to limiting factors such as finger grooves/radical shapes and esoteric operational controls that are only comfortably used in a single grip orientation.)
3. Enhancements/Coatings/Finishes
   - This is another great topic to start a fight with. There are several valid attrtutes of enhancements, coatings and finishes.
   - There are also several ridiculous myths in this category.
   - We will discuss the valid attributes.

First, there is corrosion resistance. Yes, coatings do increase corrosion–resistance, except for the cutting edge and any portion of the tool where the coating is compromised, exposing the base material. Sure, some coatings are 96-98 R/c, hardness, however they exist as a coating measured in millionths of an inch and no matter how hard, such a coating can and will be compromised in "tactical" use.

Examples of coatings range from oil and grease to paint and exotically applied thin materials that can provide lubricity and increased wear resistance if not subjected to abuse. Another factor here is the aspect of camouflage (as in the case of military application). This is done to disrupt any or all of the following: shape, shine, shadow, silhouette and movement. We consider this a very valid attribute. We included finishes here to make a final point on this topic. A bead blast finish, though useful in achieving a non-reflective effect, can be more corrosive than a polished but very reflective finish. Why? Microscopically speaking, the key phrase is "uninterrupted surface." The polished piece provides an example of an uninterrupted or slick surface where the bead blast surface is very "interrupted" or uneven, trapping moisture and magnifying other galvanic affects which all increase the likelihood of corrosion.

Much of this information is applicable and crosses over to tools outside of the edged tool category. However, in summary, this is a concept or philosophy based on a scientific approach. We here at Strider Knives Inc. hope that you will find this information useful and we invite your correspondence regarding this or related subject matter at: *striderknives@aol.com*.

Part II
Parts and Selection

Part II

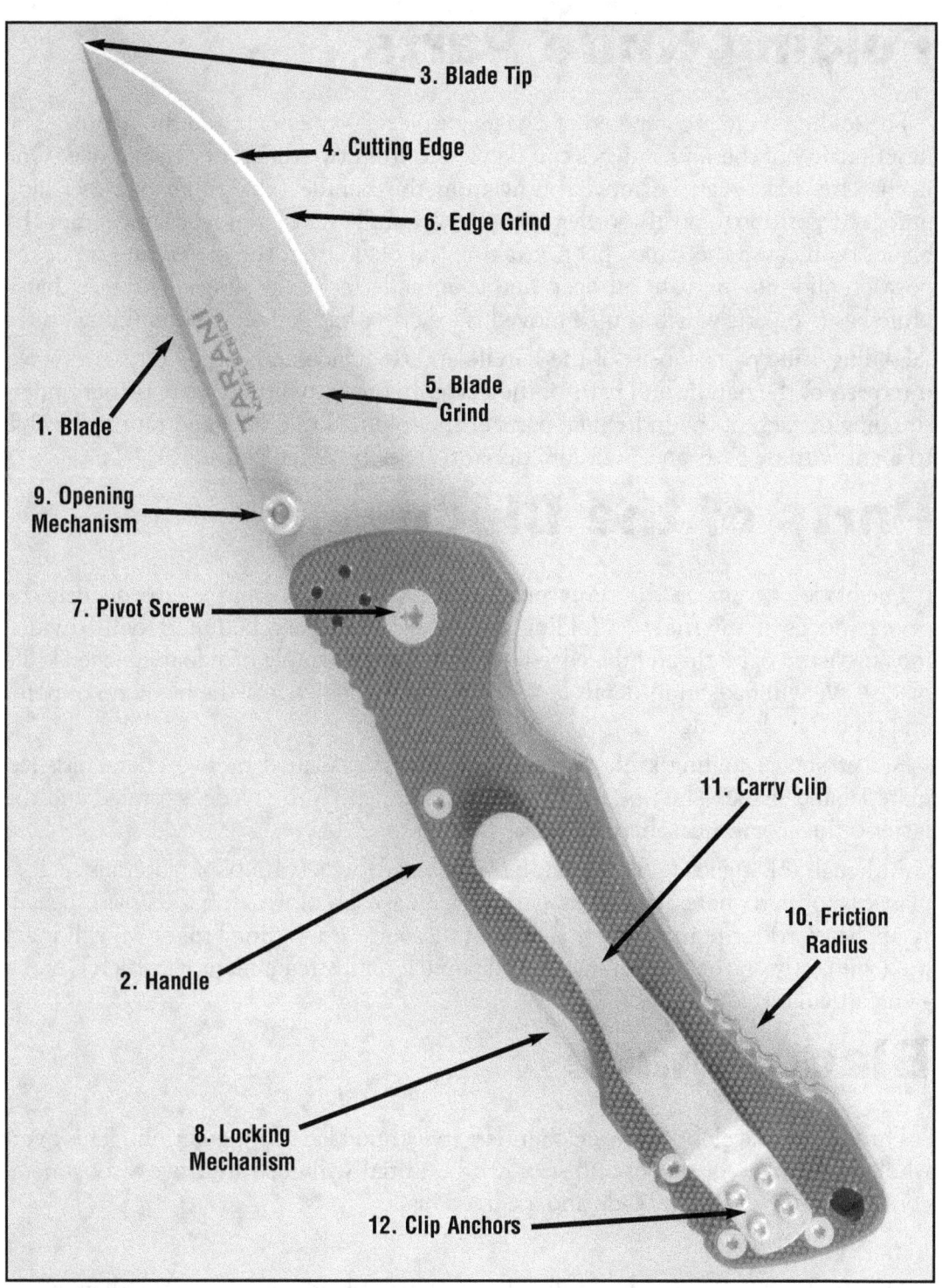

A quality folding knife and its twelve characteristic parts (each of which will be covered in detail): 1. Blade, 2. Handle, 3. Blade Tip, 4. Cutting Edge, 5. Blade Grind, 6. Edge Grind, 7. Pivot Screw, 8. Locking Mechanism, 9. Opening Mechanism, 10. Friction Radius, 11. Carry Clip, 12. Clip Anchors.

Folding Knife Parts

The folding knife is composed of two major pieces – the blade and the handle. The functionality of these two pieces can be viewed from two different perspectives. One is that the blade can be moved away from the handle (and from its closed and unlocked position) and the other is that the handle can be moved away from the blade. Both cause the same effect of taking the blade from the closed and unlocked position and moving it to an open and eventually locked position. The fact that a knife has two parts which can be moved is exactly what defines it as a folding knife.

Folding knife parts consist of the handle and the blade and can be further divided into parts of the handle and parts of the blade. In order to further increase our understanding of each of the individual parts of the folding knife it's important to be able to identify these parts and their functionality. (See full page iamge 41)

Parts of the Blade

The blade of course is the centerpiece of the folding knife and is subsequently the primary focus of any study on folding knives. All blades can be functionally divided into two parts – the tip and the edge (or edges in the example of a double-edged knife blade). We will take an in-depth look at each individually, first the blade tip or point and then the blade edge

As per above, folding knife blades are generally classified by two characteristics: material and shape. The first is the material from which the blade is formed and the latter is the geometrical shape of that material.

Although folding knife blades can be from a truly wide variety of materials such as plastics, polycarbonates, fiberglass, ceramics, etc, (each of which has its own specialty application) for purposes of this general discourse on folding knives we will focus predominantly on the most common and popular mainstream material which is metal – in particular steel.

Blade Shape

Throughout the globe there are countless styles and blade configurations. This overwhelming volume of shapes and sizes can be initially divided into two very general configurations: a straight blade and a curved blade.

Straight Blade

The straight blade is a term applied to any type of blade that generally flows along the path of a straight line. Although the very edge itself may curve slightly nearest the tip (for example a Drop Point or Spear Point tip), the overall blade shape – in its entirety - including the back of the blade (on a single edged blade this is referred to

Part II

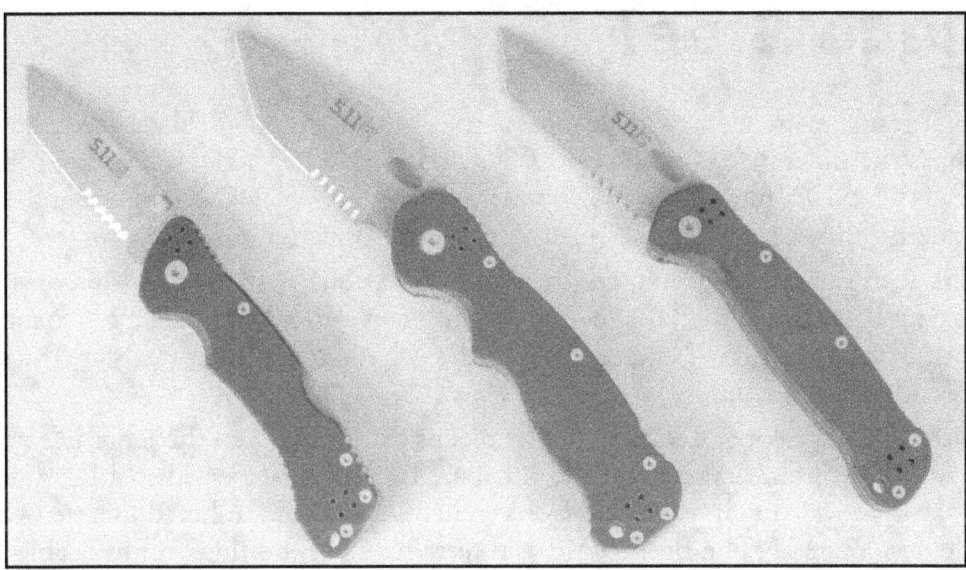

Examples of Straight Edged blades

as the "spine") remains in a relatively straight line.

Curved Blade

The Curved Blade is clearly differentiated in shape from the Straight Blade simply by physical appearance. The curved blade shape can be traced to both Indian and Arabic origins. Later on in history this shape was fused into the design of the hand-held fixed curved blades of Malaysia, the Southern Philippines and throughout the Indonesian Archipelago. Still later on down the timeline the folding version of the fixed curved blade was developed and placed into mainstream production circa 1997.

Examples of Curved Edged blades

Blade Steel

As a result of frequently asked questions in the field about blade steels such as "What's the difference between S30V and 154CM? What about 440C compared to ATS-34?" and to further expand upon the above chapter "Metallurgy and Blade Geometry" this section delves deeper in response to such inquires.

Although numerous in volume, there are only so many steels that are considered viable for usage as a blade for a modern folding knife. In making reference to modern mainstream production of folding knife blades, what differentiates steels are the many different grades of steel.

Historically speaking, steel was a fairly recent discovery given the lengthy corridors of time. The Bronze Age, generally attributed to the first period in history where metal was in wide use in varying regions (roughly 3,000BC), ushered in a new era for blade development. The Bronze Age was terminated at the time iron was able to be heated and forged which ushered in the Iron Age. As time moved on newer discoveries were able to turn iron into steel which was found to be far superior in performance and durability.

One of the most significant developments in the history of steel occurred in the middle of the 19th century when chemists and metallurgists discovered the differences in the chemical composition of different grades of steel. Some were noticed to be higher quality than others. In the mid 1890's a German metallurgist named Adolf Martens identified characteristic microscopic patterns of the hard quality constituent of which quenched steel is chiefly composed. To this very day these quality characteristic patterns are known as Martensite in honor of the founder.

The types and grades of steel utilized in folding knives warrant an entirely dedicated manuscript. As a mater of fact, Metallurgy – the science and technology of metals – as was covered earlier, is a comprehensive study all on its own. In our modern era of technological advancements in the areas of Metallurgy (including fabrication of alloys, annealing, tempering and cryogenics), blade steel types, grades and information regarding characteristics and properties could literally fill entire volumes. This vast body of information is staggering and is outside the intended scope of study for this manuscript. To stay on track with the study of folding knives, carry, selection and usage, the following is a brief rundown on the most common types and grades of steel mainly available to both production and custom folding knife manufacturers at the time of this writing.

CPM S30V

American-made CPM S30V a product of the Crucible Materials Corporation is the undisputed industry leader in folding knife steel. CPM S30V is what the author (and other contributors to this manuscript) considers to be a very high-quality knife steel.

As an expansion to the above chapter on Metallurgy and Blade Geometry and with

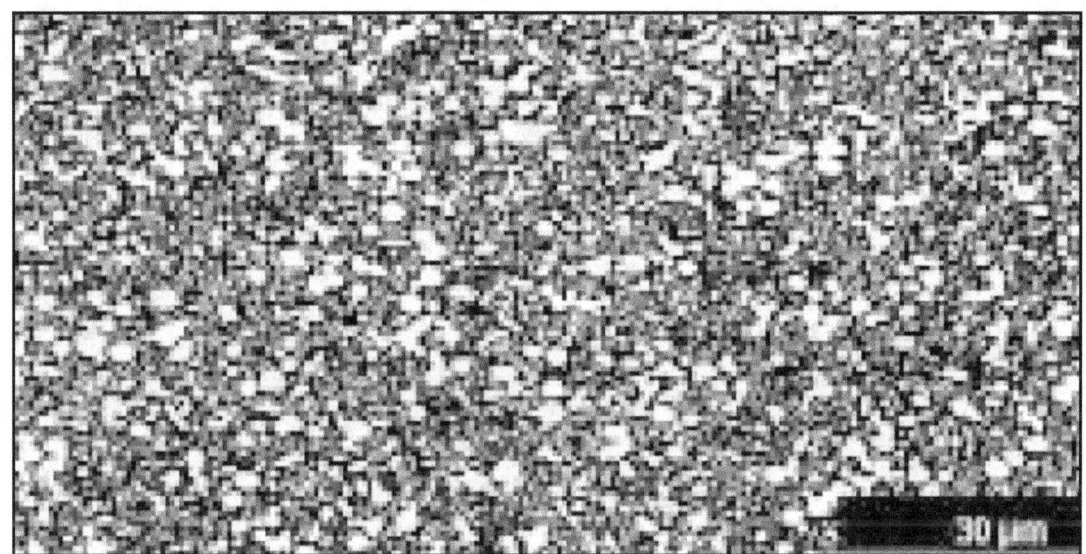

CPM S30V is made by the proprietary CPM process, which results in a homogeneous, fine grained microstructure with uniformly dispersed carbides, as can be seen in the magnified photo above. The composition of S30V is balanced to promote the formation of vanadium-rich (MC type) carbides which provide better wear resistance than chromium-rich (M7C3 type) carbides.

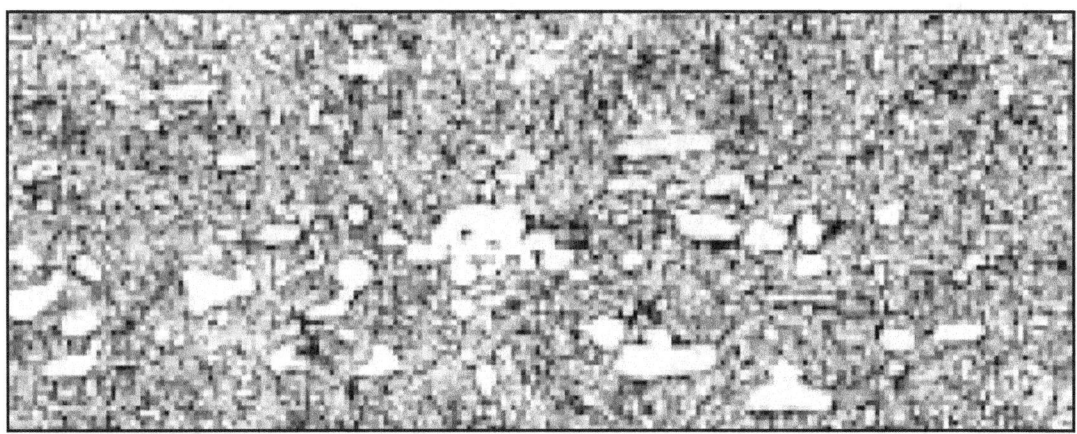

Conventional AISI 440C is a martensitic stainless steel containing chromium (M7C3 type) carbides for wear resistance. This typical 440C sheet microstructure reveals carbide banding which reduces toughness and, depending on the severity and location, can cause chipping at a very fine edge.

regards to FAQ's from the field, the following information has been adapted for this manuscript from the Crucible Materials Corporation web site and Data Sheets for CPM S30V. It provides some technical comparisons to other common high-end knife steels and can explain, to a degree, why CPM S30V excels.

CPM S30V is a martensitic (derived from Martensite as per above) stainless steel designed to offer the best combination of toughness, wear resistance and corrosion resistance. Its chemistry has been specially balanced to promote the formation of vanadium carbides which are harder and more effective than chromium carbides in providing wear resistance. CPM S30V offers substantial improvement in toughness over other high hardness steels such as 440C and D2, and its corrosion resistance is

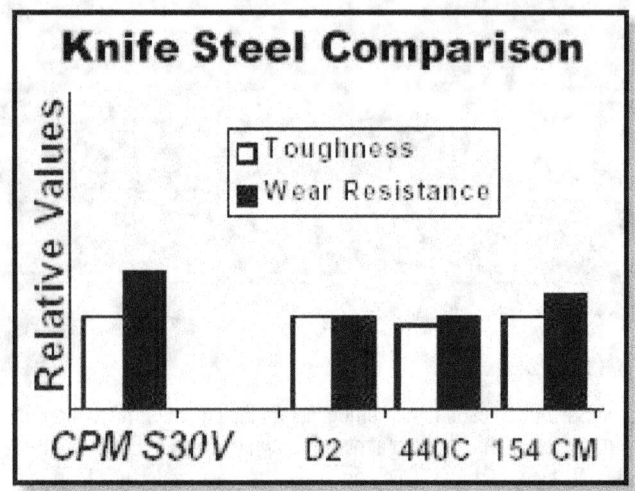

Crucible CPM S30V Alloy Composition
Carbon 1.45%
Chromium 14.00%
Vanadium 4.00%
Molybdenum 2.00%

equal to or better than 440C in various environments.

The process of producing CPM (Crucible Particle Metallurgy) steels involves gas atomization of pre-alloyed molten steel to form powder. This powder is then screened and then isostatically compressed into 100% dense compacts. The CPM process produces steels with no alloy segregation and extremely uniform carbide distribution characterized by superior dimensional stability, grindability, and toughness compared to steels produced by conventional processes

It is important for the reader to realize at this point that we're getting into the microstructure of steels and to keep in mind how this relates to the end product and the usage of this end product.

As with all things for sale, the higher the quality, the higher the cost. Same goes for automobiles and houses. If all that the knife will be used for is opening boxes, cutting twine and opening letters, then is there really the need for consideration of microstructure balanced to promote the formation of vanadium-rich (MC type) carbides which provide better wear resistance than chromium-rich (M7C3 type) carbides?

Again, it all boils down to the simple filter of, "What will the knife be used for?"

Although the longitudinal toughness for all three of these grades is

Toughness (Transverse Charpy C-notch Testing)	
Grade	Impact Energy
CPM S30V	10.0 ft. lbs.
440C	2.5 ft. lbs.
154CM	2.5 ft. lbs.

about 25-28 ft. lbs., the transverse toughness of CPM S30V is four times greater than that of 440C or 154CM. These higher transverse toughness results indicate that CPM S30V is much more resistant to chipping and breaking in applications which may encounter side loading. In knife making, its higher transverse toughness makes CPM S30V especially good for bigger blades.

Edge Retention (CATRA Testing Relative to 440C)	
Grade	%
CPM S30V	145
440C	100
154CM	120

The CATRA (Cutlery & Allied Trades Research Association) test machine performs a standard cutting operation and measures the number of silica impregnated cards which are cut (TCC = total cards cut). It is considered a measure of relative wear resistance.

Corrosion Resistance Average Pitting Potential measurements (below) from Polarization Curves run in 5% NaCl (Sodium Chloride) Solution at Room Temperature: (Higher voltage pitting potential indicates better corrosion resistance.)

In an attempt to include brief yet concise information on mainstream folding knife

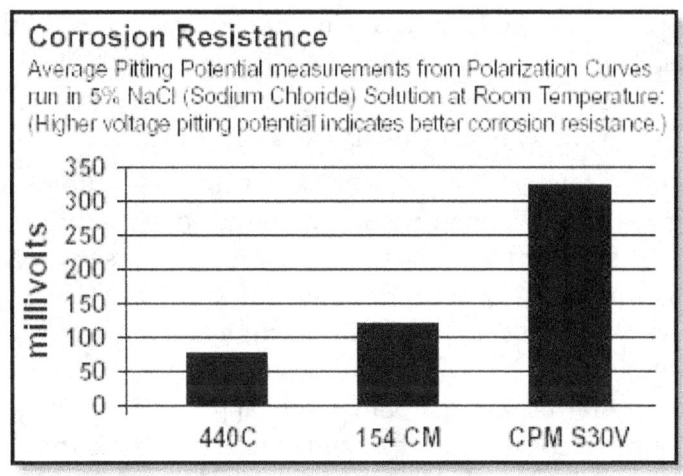

Note: Properties shown are typical values. Normal variations in chemistry, size and heat treat conditions may cause deviations from these values. Note: this information provided courtesy of Crucible Materials Company.© 2003 Crucible Materials Company. Thanks again to Mr. Dwyer for his efforts in acquiring permission from the Crucible Materials Corporation to utilize this information throughout this manuscript. For more information on CPM S30V click on http://www.crucible.com

steels for purposes of this abbreviated study, the following is an excerpt from a web page created by one Zoe Martin which in my humble opinion provides a satisfactory overview.

5160 is a common spring steel, basically 1060 with one per-cent of chromium added to make it deep hardening. (It may still be selectively drawn with a softer back, if desired.) An excellent steel for swords, or any other blade that will have to take some battering, the choice of Jim Hrisoulas who makes some of the finest working swords in the business. Long blades are best around the mid 50's on the Rockwell scale, while small, working blades can be put into service at a full 60 RC. Forged blades with a well packed edge seem to cut forever! Rough on grinding belts. Jokingly called O-C-S, old Chevy spring.

52100 is a ball bearing steel, generally not found in useful grinding sizes, but terrific in edge holding and toughness. 52100 is 5160 with an attitude, more alloy and more carbon that makes it harder and tougher. Like 5160, throws a brilliant yellow spark. Ed Fowler has developed a superior heat treating technique for this steel.

L-6 is the band or circular saw blade steel used in most lumber mills and downright hard to find in any other form. Hardens in oil to about RC 57 and takes a fine edge for most cutting, particularly where the edge might be steeled back into shape. The steel performs exceptionally well where flexibility is needed but rusts easily, like virtually all of the simple carbon steels. L-7 is the same stuff with a little more carbon.

A-2 is a type of steel, with fine wear-resisting qualities plus excellent resistance to annealing and warping. Grinding is noticeably harder than 0-1 but not extremely difficult. Sawing is tougher and relates to the five percent of chrome in this steel's chemical make up. Really nice to finish with the grinder and very little grain appearing in buffing. Excellent flexibility. Phil Hartsfield gets incredible cutting ability out of this steel. Several other of the A series will also make fine blades.

D-2 offers another air hardening tool steel, but with 12% chrome and excellent, if not superb, wear resistance. The resistance also holds true in both sawing and grinding, even while the steel is fully annealed. While using belts up at a faster rate than average, D-2 is not particularly hard to grind with fresh belts. Using old belts causes enough heat to work harden the steel. D-2 anneals at a somewhat higher temperature than A-2 and will not take a true, mirror polish. Definitely a steel for the advanced craftsman. Its major drawback is the orange peel appearance of the surface when finished to a high gloss. One knife maker is often quoted as saying that D-2 takes a lousy edge and holds it forever. Often found as surplus wood planer blades. D-4 and D-7 are also good cutlery alloys, but darn hard to find in the right sizes. Air hardening steels can work harden while you're grinding them if you get the stock too hot. This doesn't mean much on the grinder, but when you try to file a guard notch, the file will just slide.

M-2 is a high temperature steel made for lathe cutting tools, which has darn little to do with knives, but allows you to really cook the blade in finishing after heat treat without annealing it. M-2 is perhaps a bit better in edge holding than D-2. It is also rather brittle and not recommended for large knives.

440C was the first generally accepted knife makers' stainless and remains quite popular, particularly since the sub-zero process was developed to add toughness. On the grinder, it's gummy and gets hot fast, but it cuts a lot faster and easier than any of the carbon steels. Your belts will cut about 2 to 3 times as much 440-C than 0-1. Using hand hacksaws on it will wear out a lot of blades in a hurry. But with the proper care, good heat treating and finishing, 440C produces an excellent, serviceable and durable knife, even for the new knife maker. Anneals at very low temperature. Please note that 440A and 440B are similar alloys, often confused with 440C, but not worth a damn for knife making use. Commercial knife companies often mark blades 440 when they're one of the less desirable versions, giving the real stuff a bad name. 440C is also available in more sizes and in more places than just about any stainless alloy suitable for knives. It is also essential to remember that collectors hate to see one of their prizes turn brown in the sheath, and 440C resists corrosion very well. While the variation, 440-V doesn't seem to get quite as hard, but holds an edge for much longer and is much more difficult to grind.

154CM was considered by many to be super-steel, if you can find some of the old production stock. The new batches are not manufactured to the standards that we've come to expect for knife steel. While excellent in use, 154CM eats up the finest hacksaw blades in one across-the-bar cut of 1-1/2". It's machining and grinding qualities are similar to 440C and won't win it any awards for ease in working. In use though, this alloy has a definite advantage in both hardness and toughness over 440C. 154 CM is not an accepted standard grade designation, rather a manufacturer's trade name.

ATS-34 Japanese made stainless considered the equal of 154CM. Import restrictions have been eased somewhat, although they were forced to raise the price by 50%. Cleaner than the 154CM. (154CM is no longer used in government specified applications and is not the vacuum melt product that we once appreciated.) ATS-34 is virtually the exact same alloy as 154CM, minus 0.04% of one of the less essential elements. ATS is double vacuum melted and very clean. It also comes with a hard, black skin that will put a shine on your grinding belt before you know it. We recommend knocking the skin off with old belts before tapering the tang or vee grinding. One fellow tried to take the skin off with an industrial motor driven wire brush wheel. All he did was polish it. We now stock a belt specifically designed to remove this scale. ATS-34 is a trade name.

AEBL seems to be about 440B. Extremely easy to grind, in fact, I think I may have set a world record with it a few years back, over a hundred blades from bar stock to 220 grit within eight hours. Heat treat like 440C. Edge holding is best when heat treating includes a freeze cycle. Very easy to polish and buff. Very nice choice for miniatures, kitchen knives, etc. AEBL has several quirky habits in grinding that make it difficult to use on thicker or larger knives. Makes nice kitchen knives. "Hoss" uses this in his beautiful stainless Damascus and reports that it holds up very well.

420 modified stainless, has been successfully used by some commercial knife producers, but availability is not practical for the hobby knife maker (custom knives) since few custom knife makers order steel in mill rolls.

VASCO WEAR is rather expensive but very, very good in edge holding. Resists grinding very well too! You'll swear your belts have all gone dull when you try it. Do everything you have to before heat treating, because you sure aren't going to be able to do much afterward. Priced like lobster tails, when you can find it. Try Vasco-Pacific in the Los Angeles area. Vasco-Pacific uses their own series of names for their alloys.

DAMASCUS steel is such a widely made product that it is impossible to make too many general statements about it, other than it seems to catch collectors better than any other type. Each smith does his in a slightly different way, ranging from the fellow who toughs it out, starting with three layers, to the guy who welds a 300 layer sandwich of shim stock into a billet with one hit in a 40 ton press. They're all pretty. Reese Weiland suggests that the last etch of a Damascus blade be done with phosphoric acid, which will sort of, parkerize the metal and help protect it. He said that you have to play around with the concentration of the acid and immersion times a bit, depending on the steel you're using. This will also work on most carbon steel blades. If a Damascus blade has been hardened with a softer section at the spine or guard, you will get a much better looking etch if you use muriatic acid first, to get the depth you want, and then ferric chloride for adding color.

STELLITE 6-K fits into the same category as Vasco Wear in the wear resistance area, but doesn't need heat treating since there is no iron in it at all. The trick is exceptionally hard particles embedded in a rather soft alloy. Very flexible and easy to bend. Virtually cannot be brought to a mirror finish. Stellite blades are very much in demand by some collectors. The alloy best suited for knives now must be ordered from Canada and costs about a hundred bucks a pound. Part of Stellite's toughness comes from the rolling process used to form the bars. Cast Stellite is not nearly as tough.

TITANIUM is only a marginally acceptable metal for a folding knife blade. It cannot be hardened much past the mid 40's of the Rockwell C scale, and that's spring, or throwing knife territory.

As mentioned above there is more blade steel out there than you can shake a stick at and in addition to all of the above you can find even more such as 8A (AUS 8), HRC 57-58, ATS-55, CPM-10V, 754, VG-10, N690, 9Cr13CoMoV, X15 T.N and many others. For this interested in additional information on the performance of various mainstream blade steels may click on: http://www.cutleryscience.com/reviews/edge_testing_II.html

---

Note: The above abbreviated listing of blade steels is Copyright ©1997 By Blades 'N' Stuff - ALL RIGHTS RESERVED. For more information click on http://www.engnath.com/public/steel.htm

Part II

Blade Points

Since the creation of the very first folding knife, the proverbial "tip of the blade" has since held many shapes. These are too numerous to list here and in fact many have even been lost to antiquity. Over the millennia folding blade tips have undergone a truly thorough development and beta-testing phase eliminating, by trial and error, those that were proven non-functional or less-than-optimal performance (i.e. too thin, too long, too short, etc.). What we end up with here in the 21st century has been refined – again based on functionality - to a relatively common few "knife industry standards." These standard blade shapes include, but are not limited to the following.

Spear Point

The term "Spear Point" is applied to any point of a blade that is in the classical spear-shaped configuration. It was named directly after the formidable hand-held weapon of antiquity which was engineered specifically to pierce or to be used for thrusting.

Today the Spear Point allows the user optimal penetration and also prying capability – although one drawback is the narrowness of the blade located at the very tip. A spear Point configuration is identified by symmetric angles on both sides and is generally the point of choice for daggers, stilettos and double-edged fixed blades.

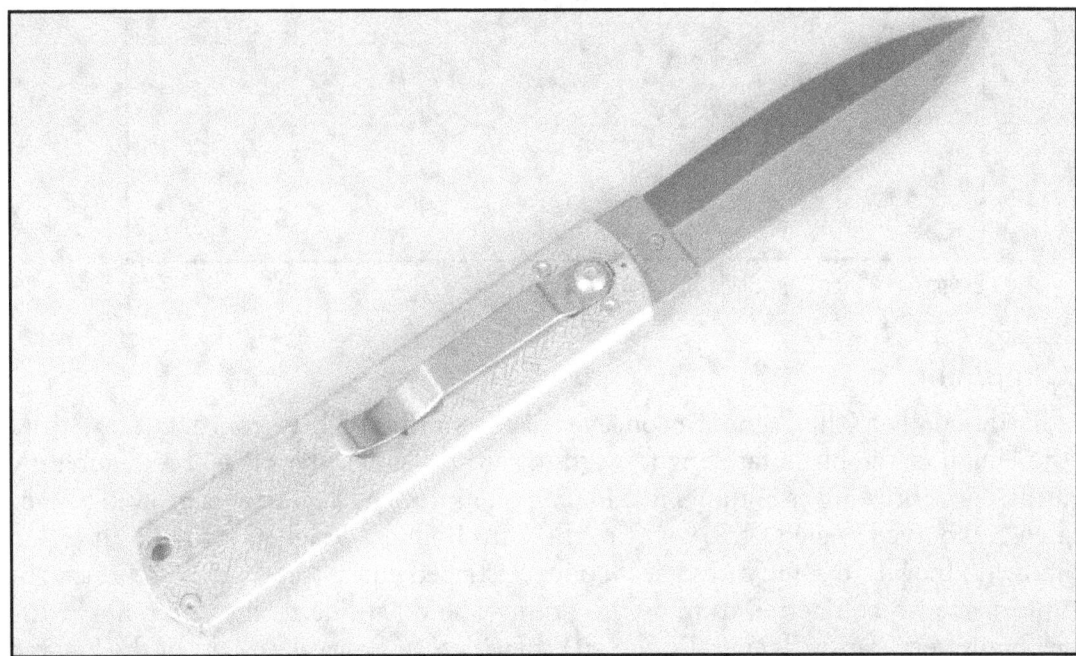

Example of a Spear Point Blade Tip

Folding Knives: Carry & Deployment

Drop Point

A slight angle on the top side of the blade tapering toward the point has the appearance of "dropping downward" – hence the name "Drop Point." Traditionally, Drop Points were designed for better handling and control of the blade and is usually found at the tip of hunting knives. Drop Points are very popular today as they provide a very strong tip and are very versatile allowing a full belly edge for optimal cutting.

It should be noted that there is a very fine line (large gray area) between where a spear point becomes a drop point. When describing certain blade shapes it can sometimes be found that these terms (spear point and drop point), based on manufacturer, can sometimes be applied interchangeably in the industry.

Example of a Drop Point Blade Tip

Clip Point

Traditionally a Clip Point is a concave shape, curving slightly from about one-third the length of the blade tapering forward toward the tip of the blade. Predominantly utilized for thrusting as it allows maximum penetration as a result of narrowed surface area and curved angle at the very tip, the Clip Point is designed for rapid insertion and withdrawal. In some cases a less-concave shaped curve (pretty close to a straight line) may also be referred to as a Clip Point. The downside to the clip point is the generally very-narrow (generally curved) tip which is the weakest part of the blade if used for prying.

Part II

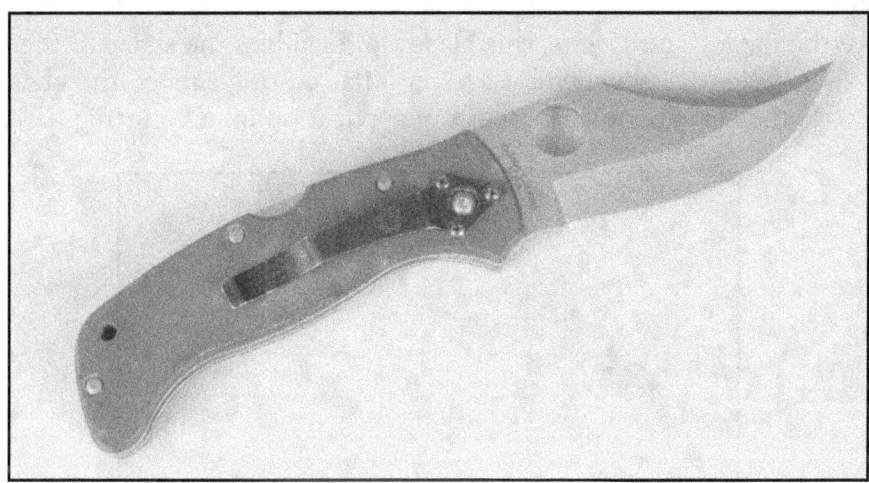

Example of a Clip Point Blade Tip

Angled Point or "Tanto" (Yoroi Toshi)

Steeped in Asian tradition, the Angle Point, often referred to as the "Tanto," is one of the most rugged blade tip designs in the industry.

The Angle or Tanto Point is characterized by hard straight angle cuts usually supported by a chisel or beveled grind and thick blade. In ancient times, the traditional Japanese Tanto was referred to as "Yoroi Toshi" – literally translating to "body-armor piercing" for which is what it was originally designed. Today the Angle or Tanto Point is synonymous with optimal thrusting and is noted for its stoutness. The thick blade tip and angled point provide very strong geometrical support for forceful thrusting and prying of rigid materials (other steels, particle board, sheetrock, etc.).

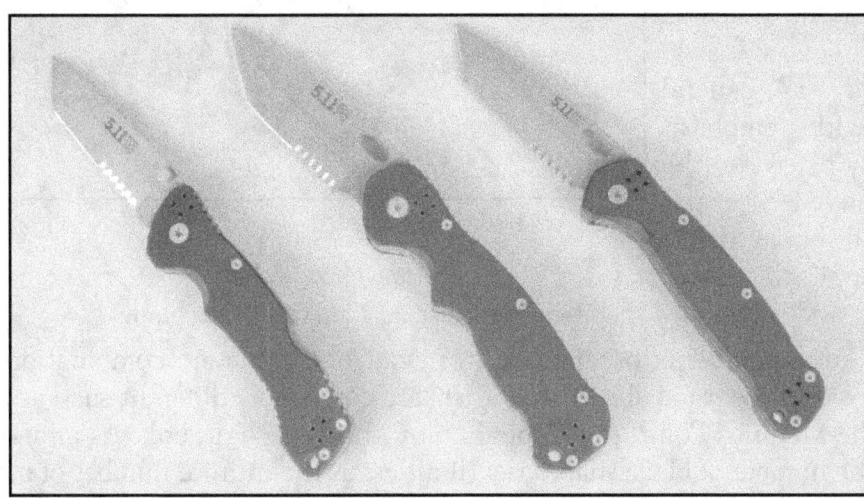

Example of an Angle Point or Tanto Blade Tip

Sheepsfoot

Just like the namesake suggests, this blade tip resembles the shape of the hoof of a sheep. It was designed for general usage on a flat cutting surface the characteristic curved top of the blade permits emphasis of application on the cutting edge.

Example of a Sheepsfoot Blade Tip

Hawkbill

Just like the namesake suggests, this blade tip resembles the shape of the bill of a hawk. Designed for general usage with emphasis obviously placed on the inside curved edge, the Hawkbill blade tip is not applicable for thrusting and is designed specifically for use as a curved knife blade.

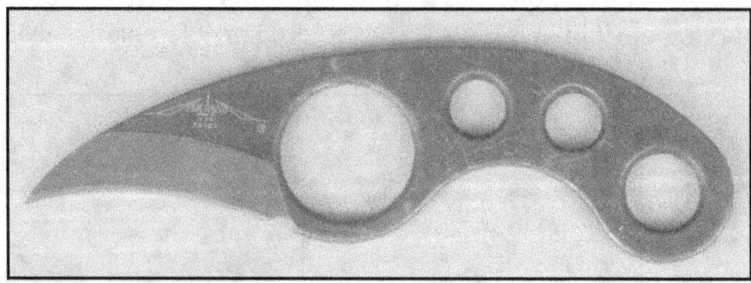

Example of a Hawkbill Blade Tip

There are a number of other blade tips available. Some are combinations and or variations of those listed above and others are completely different such as the Spey Point, the Modified Point, the Coping Point, the Pen Point, etc.. As previously discussed the number of blade shapes is as limitless as the infinite number of makes and models. Again, for purposes of our scope of study, this manuscript makes reference to the most common and popular for modern utility application.

Blade Edges

In addition to the point or tip of the blade, the other prominent feature of any folding knife blade is its cutting edge. In regards to the folding knife cutting edge, there are a few defining characteristics which should be included in any serious or comprehensive study of the modern folding knife; these are: Edge Type (single or double), Plain Edge Bevels, Serrations, Grinds and Finishes.

Single Edged

For all practical purposes (especially with regards to the professional law enforcement community, and with regards to folding knife application in the field, the Single Edged Blade is the most commonly accepted. Unlike its counterpart the Double Edged blade, the Single Edged Blade allows a higher degree of personal safety, tip and edge control and overall management of the knife. Other than for specialty applica-

Single Edged Blade

tion the Single Edged Blade is sufficient for any and all duty folding knife utility application.

Double Edged

Predominantly ground on both sides of a fixed blade, the double edged folding knife blade is not often found in the professional community. One of the main reasons is the functionality of the knife in moving the blade from its folded or unlocked position to its open or locked position and vice versa.

Other than a Balisong, Stiletto (automatic) or Trifold style folding knife handle configuration, (where the blade is fully and completely covered by the handle), it is virtually impossible to carry a conventional folding knife blade with a double edge.

First of all which one of us would like to walk around and move through our busy day with a razor sharp blade protruding along one side of the handle moving around in our pocket? Even though it would be stored in the "closed" position, obvious reasons prevent even the manufacture of such a folding knife configuration. Aside from the fact that you may not even be able to find one for sale, it's also less-than-optimal from the standpoint of safe carry and usage. A double edged folding knife for usage as a tool (aside from extreme specialty application) is not recommended by this author.

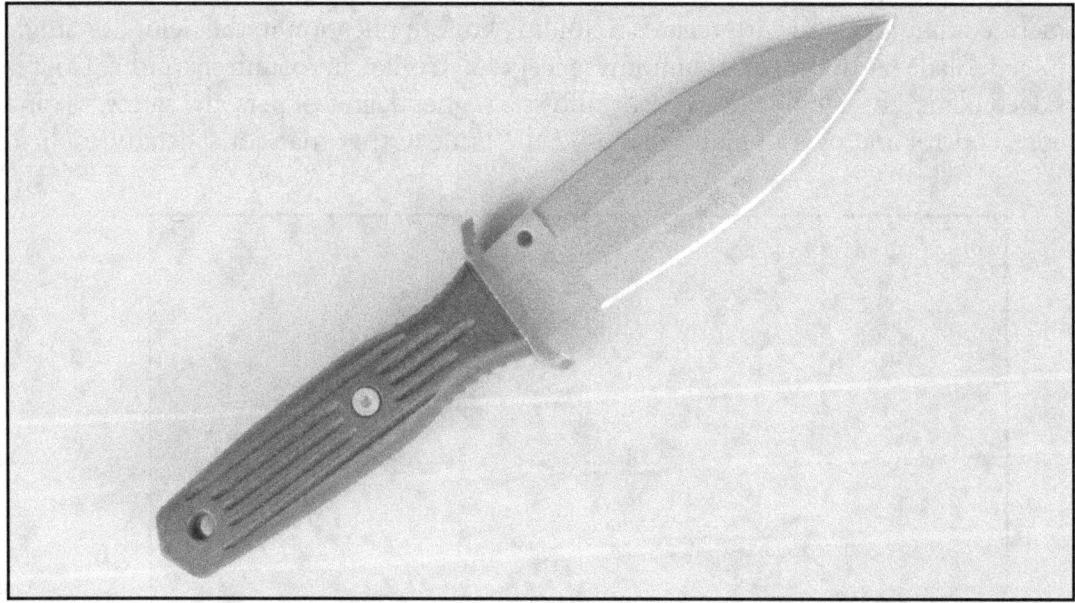

Double Edged Blade

Plain Edge

The cutting edge of any folding knife can only be one of three configurations: 1. Plain Edge, 2. Serrated Edge and 3. Half-serrated Edge or commonly referred to as "Combo" Edge. The first and most common is the Plain Edge.

The plain edged is simply defined as a smoothly sharpened edge without teeth, serrations or other modifications. A plain edge can have one of three different types of bevels – the Chisel Bevel, Single Bevel and Double Bevel.

Part II

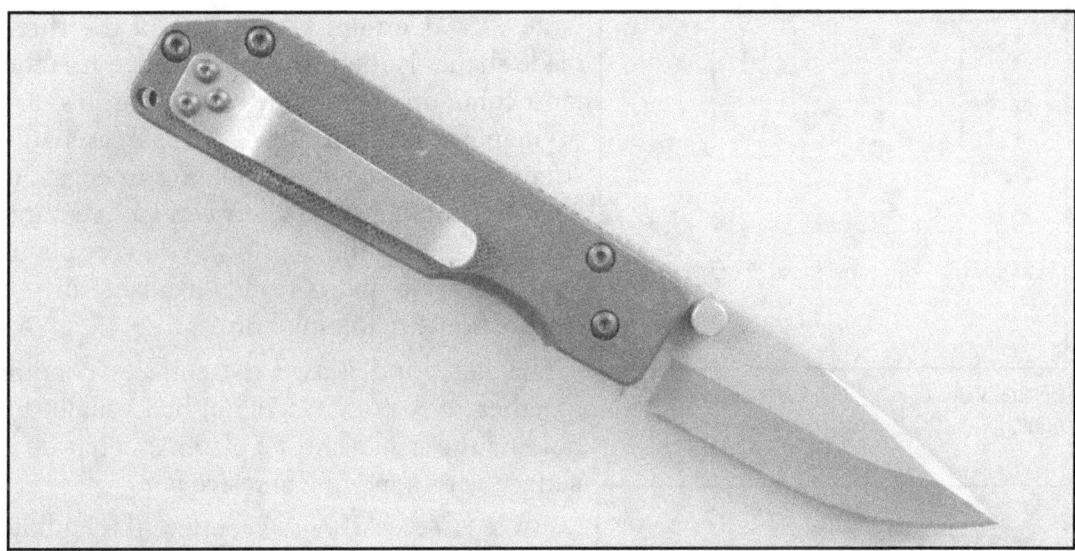
Plain Edged Blade

Blade and Edge Grinds

A blade is called a blade when it is shaped in such a manner as to move toward the sharpening of an edge or edges. As with any blade there are two grinds – the blade grind sometimes referred to as the "primary" grind and the edge grind often referred to as the "secondary grind" or "bevel." First we'll take a closer look at the primary grind and follow then at the secondary grind or actual finished edge of the knife blade.

Blade Grinds

In making a knife, the knife-maker must first grind the blade and then grind the edge.

As with any knife blade, there are only three primary shapes by which the blade may be ground. These are the Flat Grind, the Hollow Grind and the Convex grind.

Hand-grinding steel blades on a wheel.

Folding Knives: Carry & Deployment

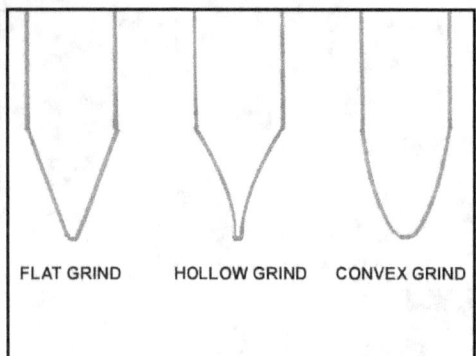

General shape of Flat, Hollow and Convex blade grinds.

Flat Grind. Certainly the easiest of the three blade shapes is the Flat Grind. Hence it is the most common of the three not only for it's easy of manufacture but also for its reasonable strength and longevity in holding an edge. As it is the "middle grind" between Convex and Concave (aka Hollow Grind) it's a very popular blade grind and is readily available in the world of quality folding knives.

The flat grind is two flat surfaces coming together to a point accomplished usually by holding the steel against a platen or a flat steel surface with a moving abrasive belt.

Hollow Grind. Without exception, the Hollow Grind (or Concave Grind) is traditionally considered optimal in allowing the sharpest edged of the three blade grinds - the reason

Hand-grinding steel blades on a belt grinder.

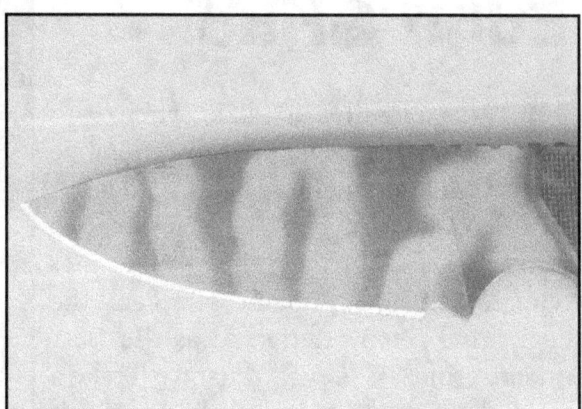

Example of a Flat Ground blade.

being a narrow cutting edge which allows maximum edge sharpness. This is usually the type of grind utilized in skinning knives. The hollow grind initiates a cut very well with its fine cutting edge. As everything has a plus and a minus, the price tag for this advantage over the other two grinds with reference to its thin fine point is that the blade is weaker. A weaker blade edge is prone to chipping, dulling and other deformities along the sharpened edge.

The Hollow or Concave Grind is usually accomplished by placement of the unfinished stock pressed against a wheel or circular steel surface with an abrasive belt or traditional "grinding wheel" moving rapidly.

Part II

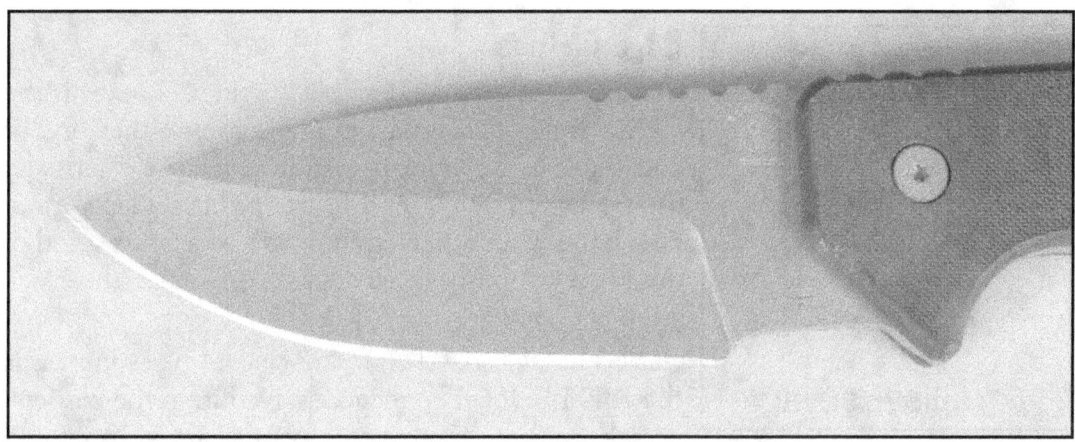

Example of a Hollow Grind blade.

Convex Grind. The third and final of the most common blade grinds is the Convex Grind. Exact opposite of the Hollow Grind, the Convex Grind (or "Appleseed Grind") is also considered the strongest of the three blade grinds given the thickness of the blade at the edge. The advantage of this is of course strength but the disadvantages are inability to thinly slice (although ideal for chopping) and difficulty in sharpening without the right sharpening gear and of course experience and knowledge in how to sharpen this blade grind. General application for this type of grind would be more likely for chopping say such as an axe or maul.

The Convex or Appleseed Grind is usually accomplished by placement of the unfinished stock pressed against a wheel or flat steel grinding surface with an abrasive belt and manipulated by hand to achieve the end desired result shape.

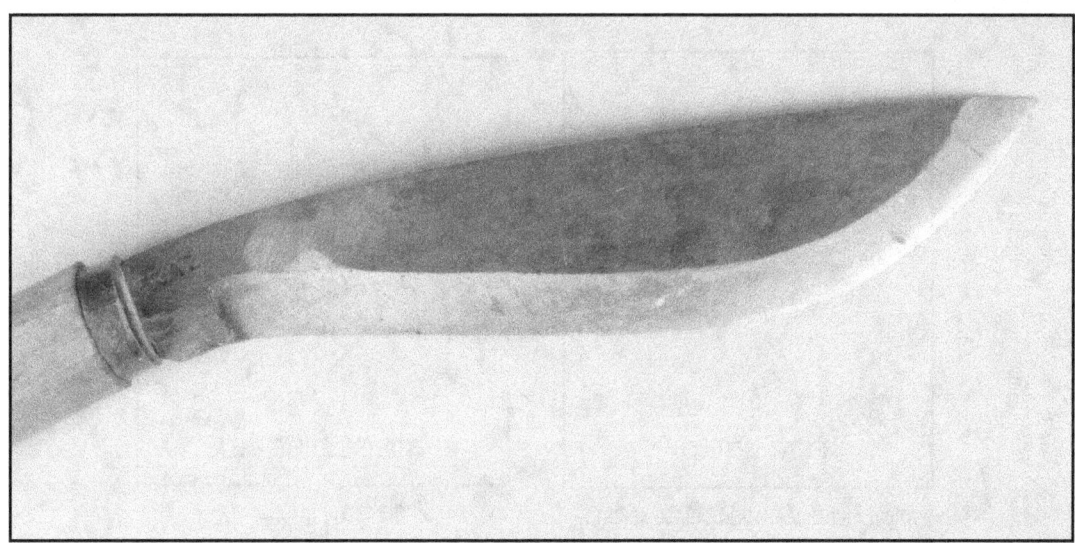

Example of a Convex Ground blade.

Edge Grinds

Considered a "secondary blade grind," a bevel is the taking away of material on angle from either one or both sides of the primary grind. Bevels are essentially what happens at the end of the primary blade grind. Bevels are what are referred to as a secondary grind of the blade as they can be applied only after the blade grind.

Chisel Bevel. One of the oldest bevel types known is the chisel bevel. The name, derived from the classical chisel shape, is applied to any edge where only one side is beveled. Sometimes referred to in the industry as a "One-sided" or "Zero-ground," this type of edge grind typically has no secondary bevel. Such edged tools as the wood chisel (and, in some cases, fine stone chisel) as well as certain traditional blades – predominantly early dynasty Japanese Katana and Wakazashi.

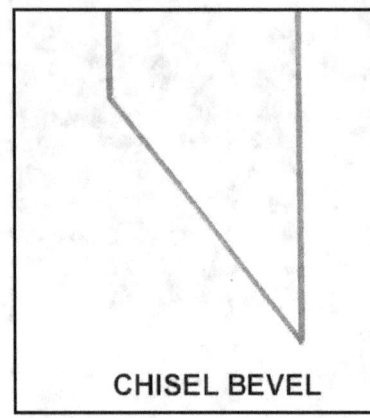
Schematic of typical Chisel Bevel.

The chisel bevel is almost always applied to a flat grind and provides the user with a much simpler sharpening solution as well as a knife blade edge with less drag points. One of the most common blades to utilize the Chisel Bevel is the Tanto or Angle Point.

Single Bevel. The Single Bevel is the taking away of material from the bottom of the primary grind only on one side of the blade. This is routine practice with various plain edges, but a mandatory for any serrated edge (full or combination edge) as it is not possible to cut serrations on both sides of a blade. All serrated blades are Single Bevel secondary grinds.

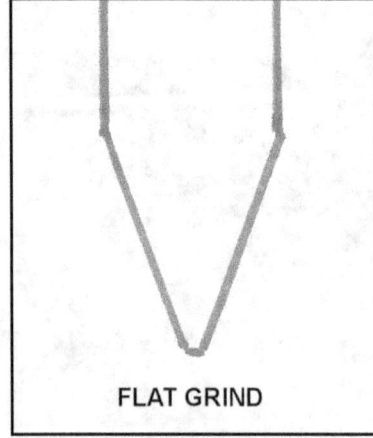
Typical Flat Blade Grind before secondary grind of Single Bevel

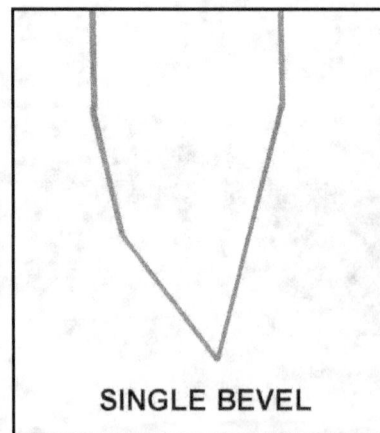
Typical Single Bevel grind.

Double Bevel. A Double Bevel is simply the grinding of two beveled edges on each side of the blade. Varying schools of thought support this type of grind as stronger than the single bevel due to the thickness of the edge. However, the Double Bevel has been known to be more difficult to sharpen.

The double bevel is the most typical and most commonly seen edge grind in knife-making.

The advantages and disadvantages of the three bevels are that the Chisel Bevel offers a very fine cutting edge that holds up very well as long as it doesn't encounter severe impact or other objects made out of the same material and hardness. The Single bevel is typically utilized for serrations with the other side remaining a sheer flat face. The "V"-grind (as it is sometimes called) or Double Bevel is the most common and is typically ground on plain edge blades (non-serrated). This grind is generally found on most production knives due to the manufacturing machinery that is involved which has two opposing wheels that the edge is drawn through leaving a double bevel or edge on both sides of the finished blade grind.

Typical Flat Blade Grind before secondary grind of Double Bevel

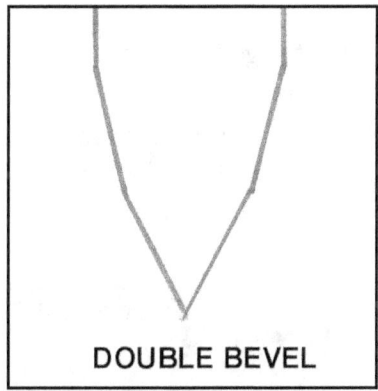
Typical Double Bevel grind.

Serrations

As time marched on, knife-users noticed that certain materials were easily cut by the Plain Edge and other materials, regardless of edge bevel, were not as easily cut. Experiments in edge grinds ensued and whether by incident or accident, it was discovered that by adding serrations to the edge there was a measurable increased functionality when the edge was applied to those materials which did not react so readily to the plain cutting edge.

A serrated knife edge is one that has "teeth" ground into it in such a manner as to form a structured row of points and recesses. Very much resembling the teeth of a saw blade, serrations function similarly with the recesses increasing the actual cutting surface of the sharpened edge and the teeth penetrating the surface of whatever is being cut with the teeth further acting as a buffer from the recesses from dulling in the process. As we will cover later in our study (see Safe Handling) varying materials respond differently based on the type of edge applied.

Folding Knives: Carry & Deployment

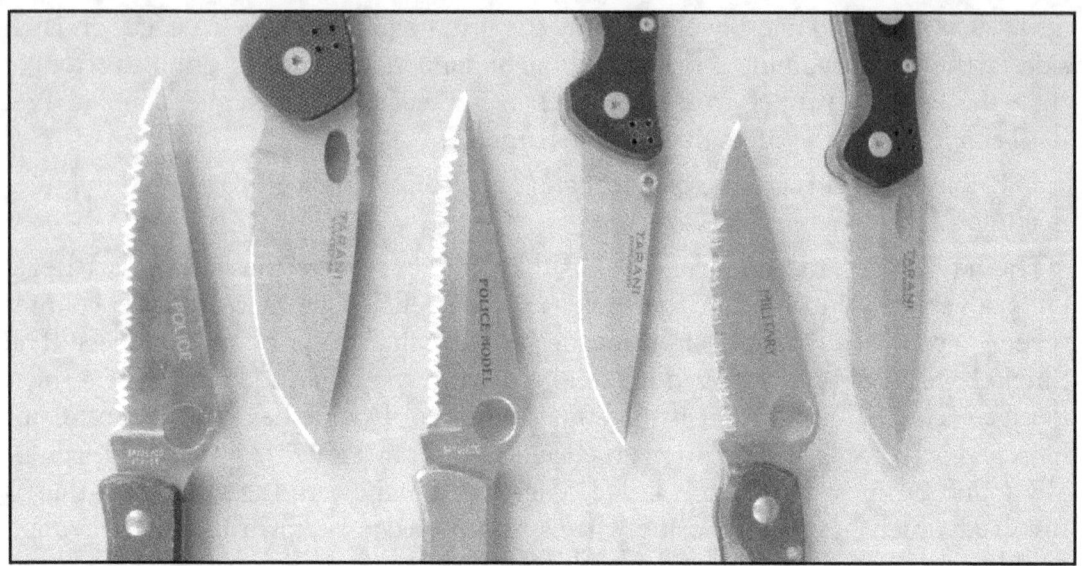

Various serrated edges illustrating front and back of this secondary grind.

In the world of edged tools there are two types of serrated edges – Full Serration and Half-serration (commonly referred to as "Combo" edge).

Full Serration. The term full serration applies to any knife blade edge that is serrated from the tip of the blade to the base of the blade. In other words there is no Plain Edge surface. The entire cutting edge is ground with these saw-teeth and recesses as described earlier.

Examples of a fully serrated blade edges

Half Serration. Commonly referred to as a "Combo Edge," the half-serration of a blade edge is a very popular cutting edge configuration. The reason for its popularity is its versatile functionality in which the attributes and characteristics of both the Plain Edge and the Serrated Edge are equally and simultaneously available to the user.

Part II

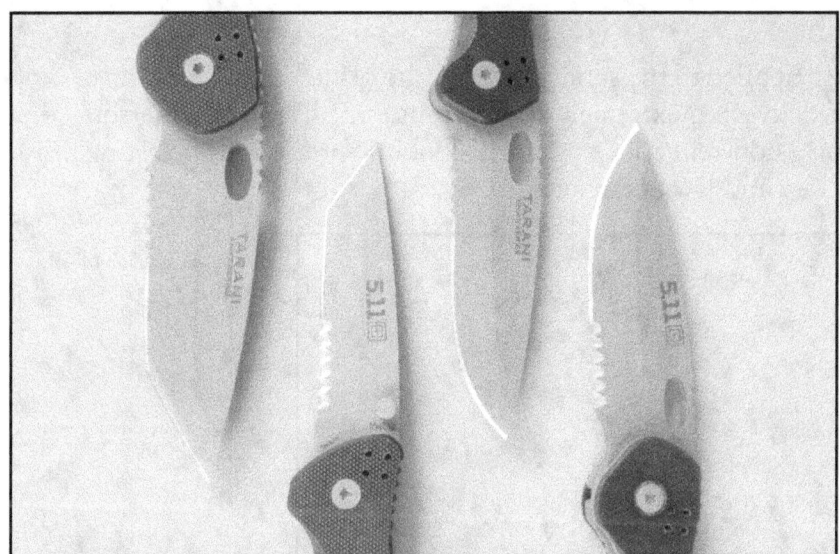

Half-serration or "Combo" blade edge. Notice both front and back of half-serration grind.

Blade Finishes

Very similar to an automobile or a house, a knife blade can be "both protected and dressed" in a particular finish. The purpose of any finish is to protect the steel from any form of corrosion or decay. A blade finish refers to the chemical treatment of the blade (sometimes on the sub-molecular level) and in some cases the manner in which the steel is finally processed so as to best preserve the quality of the blade surface.

Blade finishes are also sometimes referred to as "blade enhancements." An example of a blade finish or blade enhancement is "Blueing" which several of us are familiar with especially in regards to firearms as it is typically applied to carbon-steel alloys in order to resist corrosion and also makes an aesthetically pleasing product finish.

Other finishes that are used in the machine-tool industry such as nitride, titanium, aluminum, etc., and other such coatings enhance the ability of that tool to actually perform its function. They both dissipate heat and provide a slicker working surface.

Still other examples of how blade finishes or enhancements may be used in the edged tool industry would be as a non-reflective coating. These could be referred to as "camouflage." It can be the case that an edged tool finish may have the requirement of camouflaging a pattern in such a manner as to break up shape, shine, shadow, silhouette or movement. These would also increase the margin of success if the user of that tool has camouflage as the goal.

In the ever-expanding world of the manufacturing process of production knives, there is a correspondingly ever-growing variety of finishes and blade enhancements in which the knife blade is processed. Some of these finishes include, but are certainly not limited to Bead Blast, Coating, Polished and Satin Finish.

Bead Blast Finish

An anti-glare finish applied to the surface of the knife blade by sand blasting or bead blasting (process of delivering extremely high pressure media against the surface of the blade) producing a non-reflective and rough-looking surface. Bead blasting finish is popular with military and law enforcement.

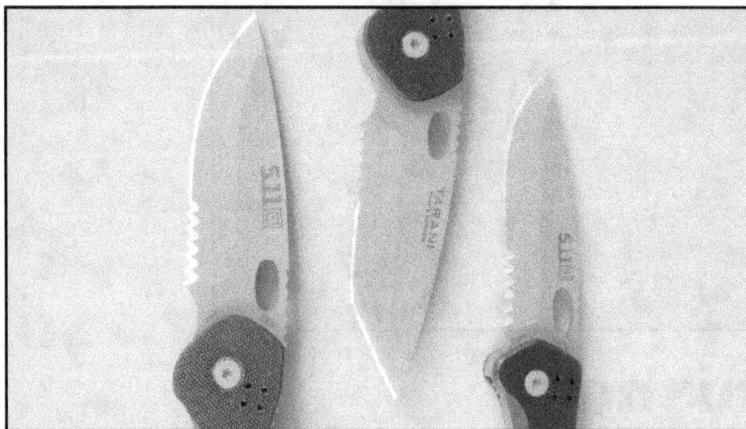

Examples of Bead Blast Finish

Coating Finish

Similar to the coating finish of gunmetal to prevent corrosion, there are a variety of coating finishes for knives. Generally, production knife companies prefer to meet or exceed various industry standards or specifications (a commonly-referred to specification with regards to anti-corrosion of steel would be the likes of ASTM-117) with regards to these modern-day finishes. Folding knives with a coating finish including a number of modern coating finish products such as BK1®, BC1, BT2, Microtome, Black Oxide, Xylan ®, Titanium Carbo-Nitride (TICN), etc., are readily available. More traditional finishes would be say Parkerization or even "blued" (really old school).

Examples of a Coating Finish.

Polished Finish

A polished finish is a high-gloss rich-luster finish applied to the surface of the knife blade which aids in corrosion resistance. Similar to the highly reflective finish of any polished metal, the knife blade surface is left with a glossy finish.

Examples of Polished blades

Satin Finish

Intended more for the knife collector and even more so on the custom knife side of the house, a satin finish provides a flat sheen to the blade surface. There are a number of varying types of satin finish and these can include Burnishing, Hand Rubbed (a type of satin finish typically applied by custom knife makers), Stonewash, or Tumbled.

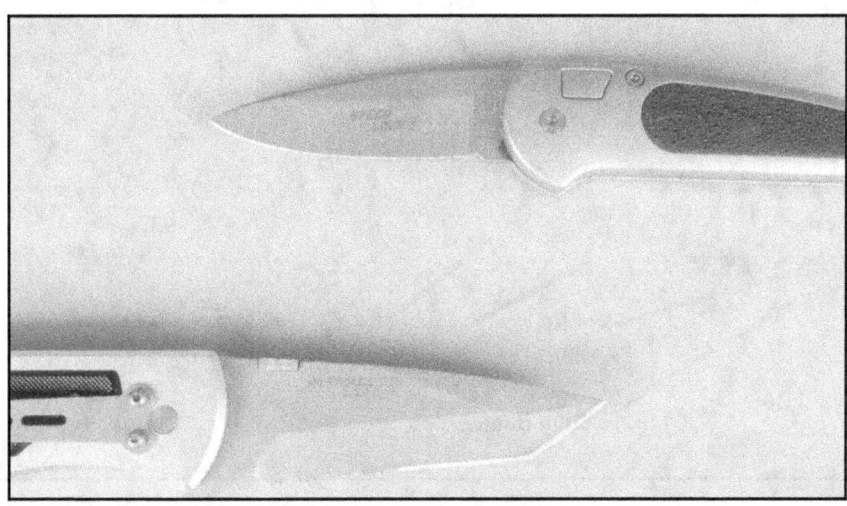
Examples of Satin Finish Blades.

Folding Knives: Carry & Deployment

Camouflage Finish

Requested by various military and specialty application, a camouflage finish is ideal to meet certain operational requirements.

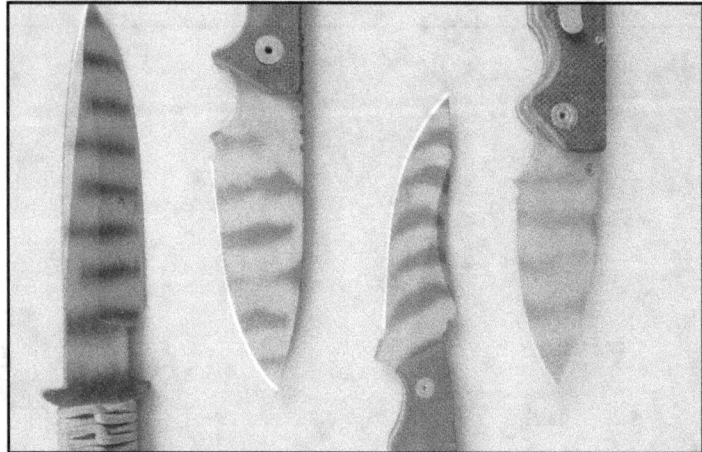

Examples of Camouflage

Parts of the Handle

In this section we will focus our study on to the remaining primary piece of the folding knife – the handle. In order to further increase our understanding of each of the individual parts of the handle it's important to be able to identify each of these parts and their corresponding functionality.

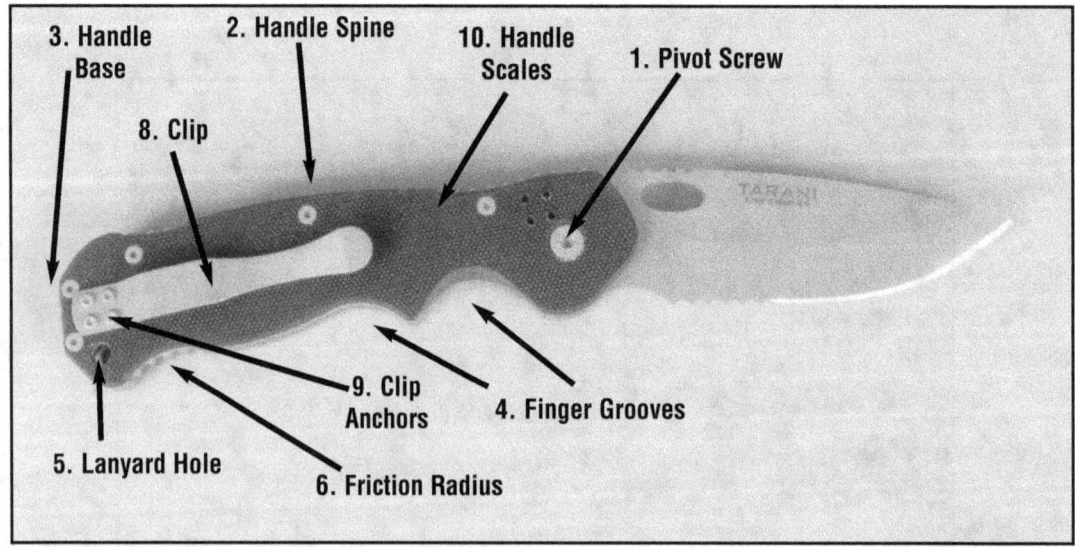

1. Pivot Screw, 2. Spine, 3. Base, 4. Finger Grooves, 5. Lanyard Hole, 6. Friction Radius, 7. Finger Cutout, 8. Clip, 9. Clip Anchors, 10. Scales

Part II

Friction Radius

Especially important in later chapters, the grip of a folding knife is paramount. One must keep in mind that the typical folding knife blade is smaller than the typical fixed blade. It is a general rule of thumb that an edged tool that is relatively smaller in size is more difficult to hold on to (grip) and control than an edged tool that is larger in size when using to cut along the edge or push or pry with the tip. Try to lift a door off its hinges with a small folding knife and then try that same feat with a large fixed blade – you will clearly notice the difference. If you will recall our earlier conversation of "everything has a price" – the advantage of a small and compact folding knife is convenience of carry in confined areas (seated in a car, attending business meetings, etc.) whereas its counterpart – a large fixed blade is difficult to carry (because of the weight) and inconvenient to wear in a confined area (as well as attend business meetings).

The folding knife handle, In order to make up for its diminished size must provide some type of additional support to the grip. This is accomplished with the aid of what is commonly called a "friction radius," "friction points," "friction ramps," "knurling," "friction pads" or "finger stops."

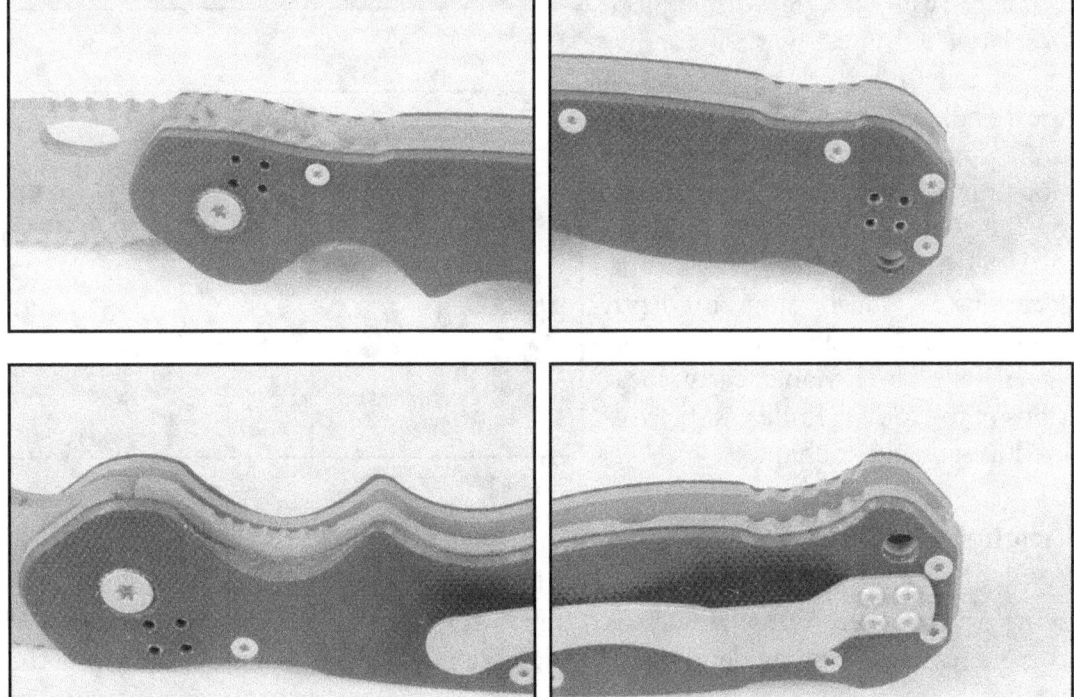

Examples of four "friction radius" positions on the handle which can be used to increase effectiveness of the grip for utilizing both the tip and the cutting edge. These provide the operator maximum protection against slippage and mishandling of the knife under pressure, inclement weather and adverse substance contact such as water, oil, or other viscous materials.

Folding Knives: Carry & Deployment

Finger Cutout

The most prominent friction radius on any folding knife utilizing a liner lock (which will be covered later under "Locking Mechanisms") is the "Finger Cutout." Housing both liner lock and the friction radius machined out of the liner itself, the finger cutout serves a dual purpose. Its primary role is of course to facilitate ease and accessibility to engage and disengage the liner lock. Whereas its secondary role is to provide a significant fore-end friction radius by which to provide additional grip support.

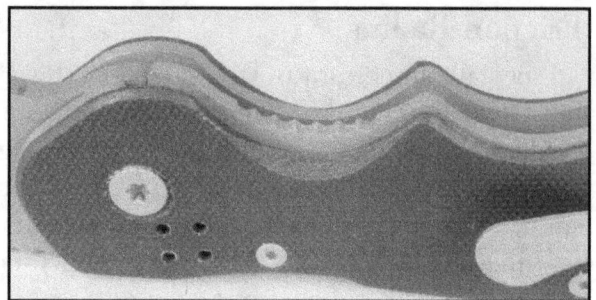
Finger Cutout

Carry Clip

The carry clip is a form-fitted piece of relatively thin yet hardened spring steel fixed to one of the outsides of the handle. A carry clip allows the knife owner to carry the folding knife not only in the pocket but actually on the pocket as well as on or along the waistband and other viable carry locations. A much more in depth study of the carry clip with regards to accessibility and viable carry locations of the folding knife will be covered in subsequent chapters.

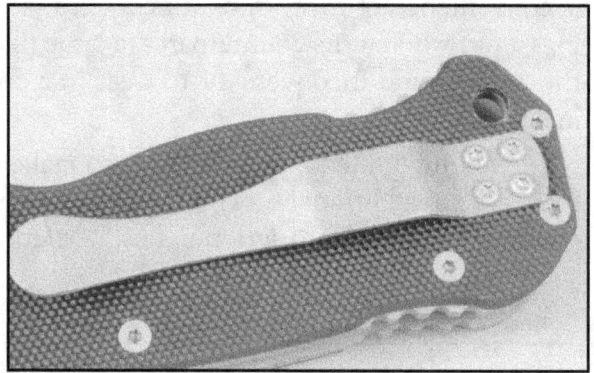
Example of carry clip.

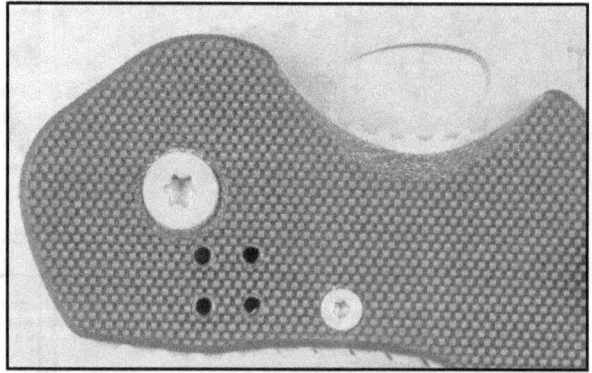
Carry Clip Anchor Holes

Anchor Holes

All carry clips are attached to the handle via anchors. The anchors are generally some type of fastening device such as a threaded screw or hook or some combination thereof (covered in more depth later) which are installed or "anchored" utilizing the anchor holes. Subsurface of the anchor holes is usually some type of threading or equivalent fastening technology based on specific clip configuration. A number of production folding knives provide multiple Anchor Hole locations to facilitate varied carry location options. Folding knife carry positions including accessibility and viable carry locations is covered in Part III.

Lanyard Hole

The purpose of a lanyard hole is of course to keep a lanyard in place. What is a lanyard? A lanyard is defined as a cord or a strap which is employed in such a manner as to hold something. In the case of a folding knife, a lanyard can be utilized to additionally assist in the access from carry position as well as used for anchoring to a specific carry position. The lanyard hole is simply the hole through which the lanyard is thread.

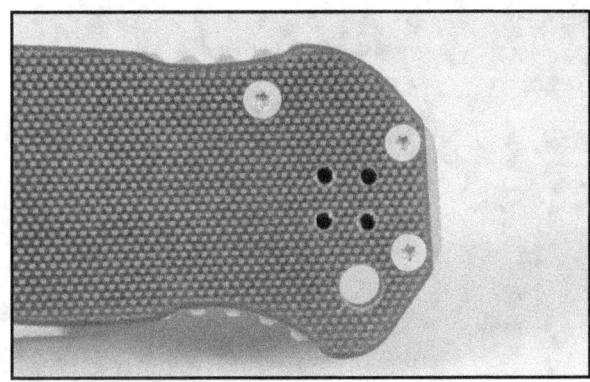
Typical lanyard can be threaded through the lanyard hole.

In today's modern production knife area, the average lanyard hole is made large enough to perfectly fit a piece of 550 cord or Paracord.

Handle Materials

In all honesty throughout my greater than 25 years in the knife industry I've seen knife handles made out of just about anything you can imagine. The quality of these materials can range from animal bone all the way up to and including G10.

As with the case of most mainstream production folding knives there are two parts to the handle where one panel mounted or fixed on either side of the closed or unlocked blade, which is also referred to as "scales."

Modern mainstream production folding knife scale materials are usually fabricated out of such modern high-tech industrial materials as 6AL4V Titanium, 300 Series Stainless steel, G10 (a carbon-filled glass reinforced laminate) and Glass-fiber Reinforced Plastic (Zytel, FNR, etc.) among others.

What should one look for in a scale? Again, as mentioned earlier form fits function and the most important and recurring question is "What will the knife be used for?" If it is to be used underwater or near and around corrosive oils or other chemicals then you'd probably want the scales on that particular folding knife to be made out of something incredibly durable such as G10 or Cryogenic G10 to withstand such harsh and abusive working environments. However, at the opposite end of the spectrum, if the intended usage of the knife is personal pocket knife for usage on the occasional cardboard box, or as a letter opener or maybe to cut the occasional string or tag from a new piece of clothing, then why pay for the additional cost of a higher-grade scale material?

Handle manufacturing process.

In the world of scales there are basically two forms of manufacture: injected molding (those scales or in some cases the entire handle that is created by the process of injecting a liquid material into a hardened steel mold and then allowed to cool) and non-injected molding, which can account for anything from laser cut to waterjet cut, to wire cut, to stamping. All of these modern industrial processes (and more) account for the majority of handles or scales used in the production of folding knives manufactured at the time of this writing.

Opening Mechanisms

As with all parts of the folding knife, the most common thread between all folding knives (except for the fact that they are composed of a blade and a handle) is variety.

An opening mechanism can be defined as any mechanism or device which may be used in such a manner as to physically move either the blade away from the handle or the handle away from the blade (depending upon your perspective), and from the closed and unlocked position toward the open and locked position.

Part II

A plethora of opening mechanisms is readily available out there in the knife market today. In the case of selecting an opening mechanism this really falls under the heading of personal preference. It's all about ergo dynamics. What fits you the best? Do you have big thumbs or little thumbs? Do you like the way a pin looks as opposed to a T-Stop? Does a hole machined into the blade feel more comfortable than an indent? Do you like ovals more than circles or circles more than ovals or neither one? Again, this particular part of the knife is truly left up to buyer subjectivity.

In order to present at least a cursory introduction to the vast array of opening mechanisms available in folding knives, an abbreviated list of such mainstream devices has been provided. The list includes, but is not limited to: "T"-stops, Pins, Posts, Waves, Holes, Indents and Grooves.

"T"-Stops

One of the most popular of opening mechanisms is the "T"-Stop as it requires very little effort to slide your thumb across the top part of the closed blade and have it catch against the 90 degree stop which in turn converts the strength of the thumb muscle into moving the blade.

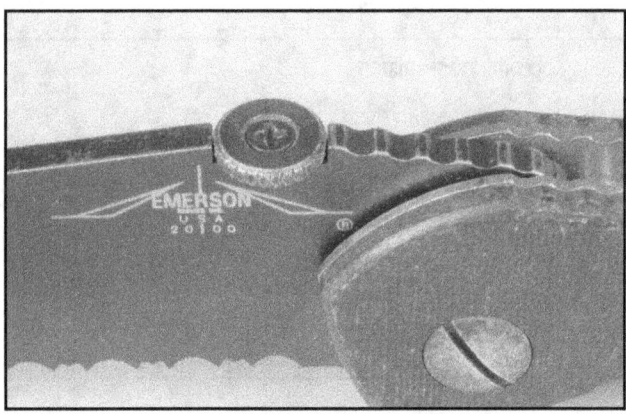

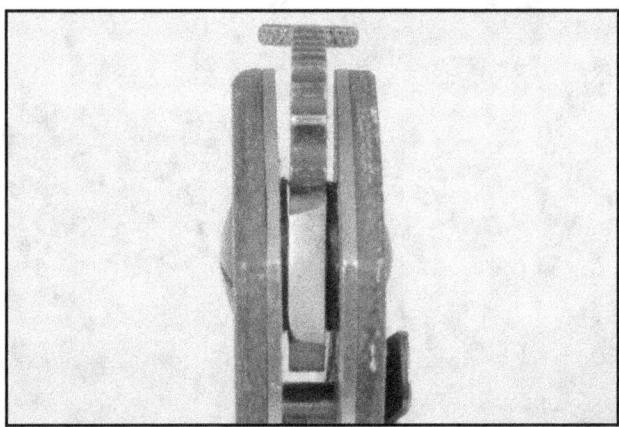

Example of a "T"-Stop opening mechanism. Note "T" shape in last photo.

Folding Knives: Carry & Deployment

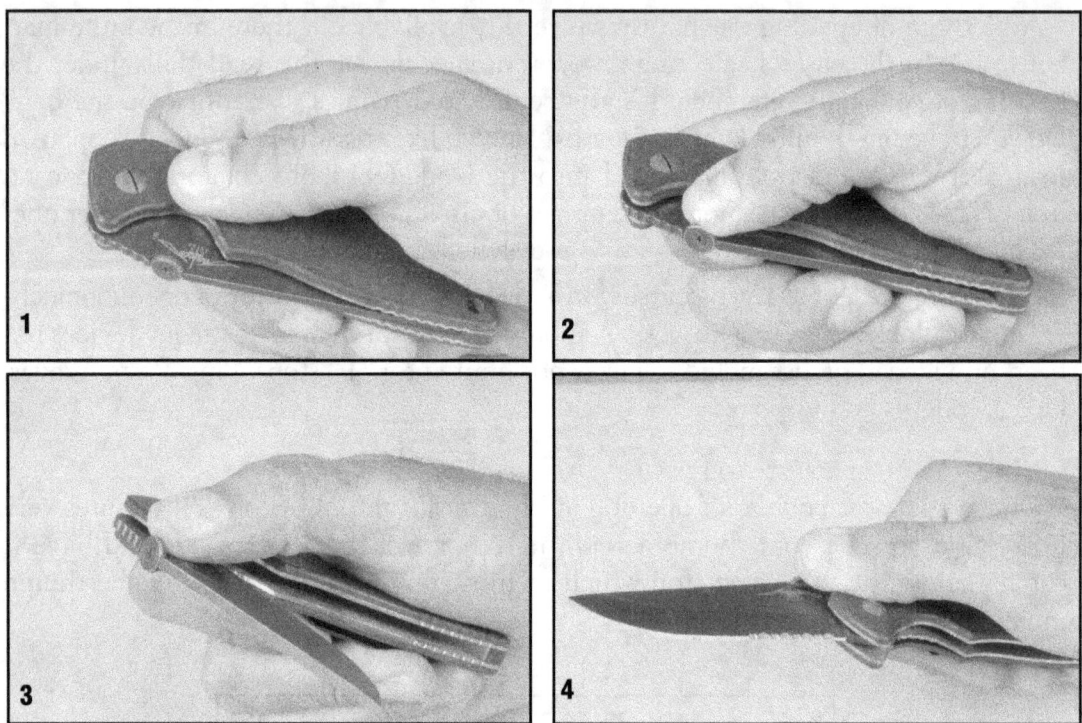

Example of utilizing a 'T'-Stop opening mechanism.

Pins

Some folks prefer a thin cylindrical "pin" that protrudes from the blade nearer to the base and back in a little from the top part of the blade. Again, there's no such case as "this is better than that" – especially with regards to Pins and Posts, it's simply a matter of personal preference.

Example of a Pin opening mechanism.

Part II

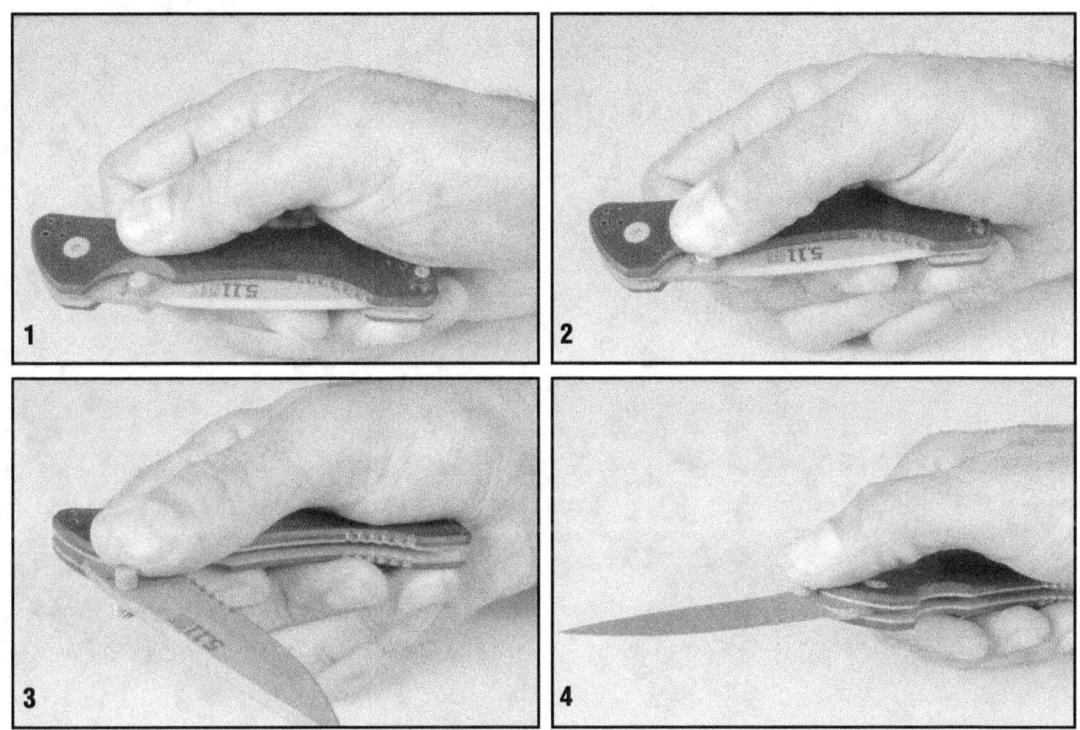

Example of utilizing a Pin opening mechanism.

Posts

Similar to a Pin, the Post is a bit thicker and provides more surface area for larger thumbs than the pin, but again – one is not necessarily "better" than the other as it boils down again to personal preference.

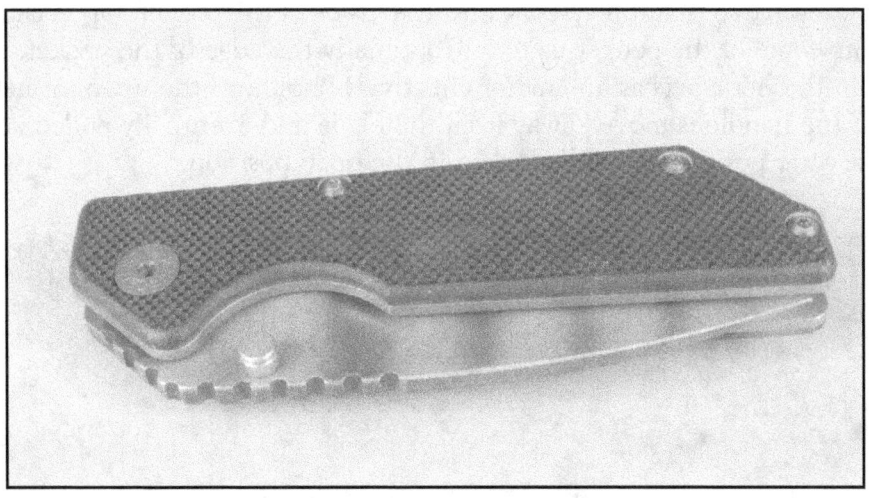

Example of a Post opening mechanism.

73

Folding Knives: Carry & Deployment

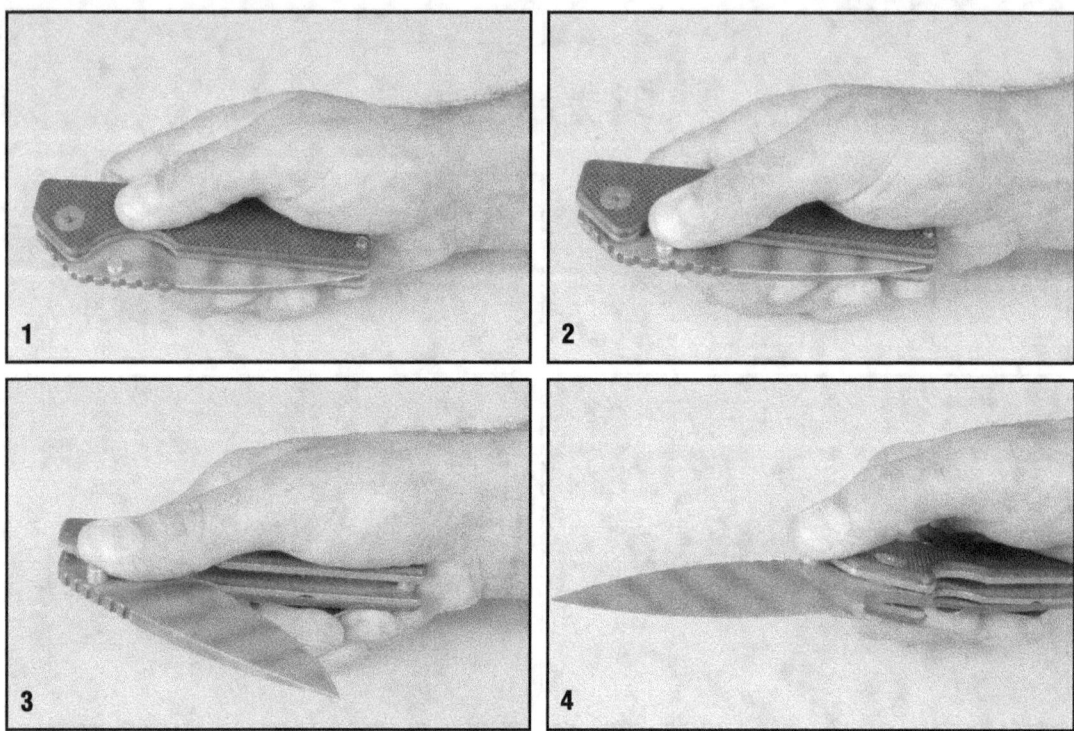

Example of utilizing a Post opening mechanism.

Wave

The Wave Opening Mechanism – or better known as the Emerson Wave ® is analogous to Emerson Knives. The wave was invented (and is patented by) Mr. Ernie Emerson – founder of Emerson Knives. The wave itself is simply a soft hook cutout at the top of the blade which faces outward when stored in a pocket. Upon pulling the folding knife (in its closed position) from the pocket, care is taken to press firmly outward against the wave so that it presses against the inside of the pocket. Upon catching the hook into some of the pocket material (optimally the edge of the pocket, the fabric caught in the hook acts as an anchor effectively "holding" the wave in such a manner that the handle is moved away from the blade and eventually pulled completely from the carry position and fully locked in the open position.

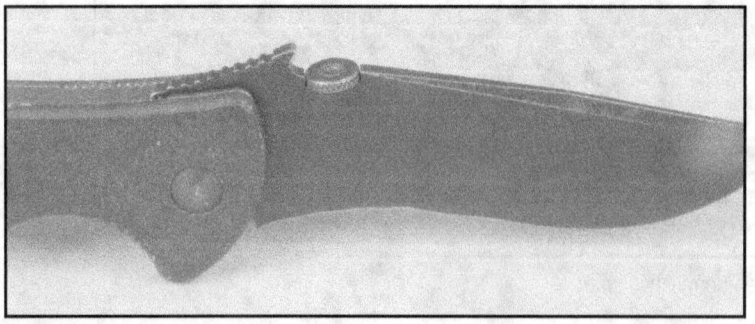

Emerson Wave® located adjacent to "T"-Stop.

Part II

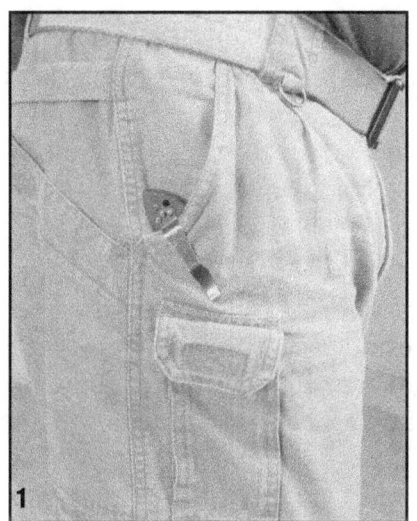 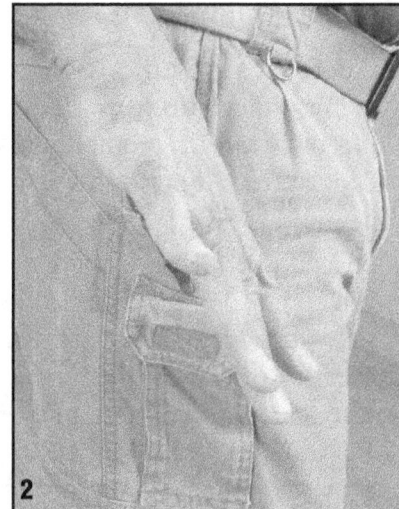

Step 1. Position the folding knife clipped to pocket opening.

Step 2. Reach for secured knife in carry position.

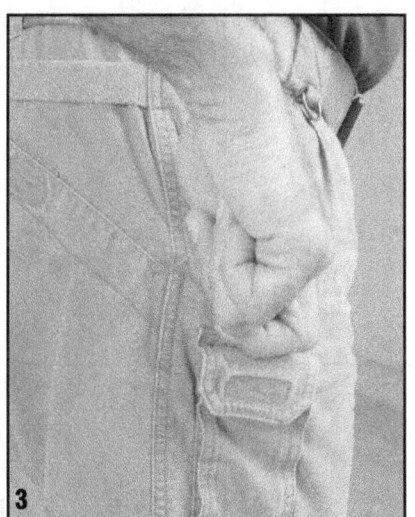 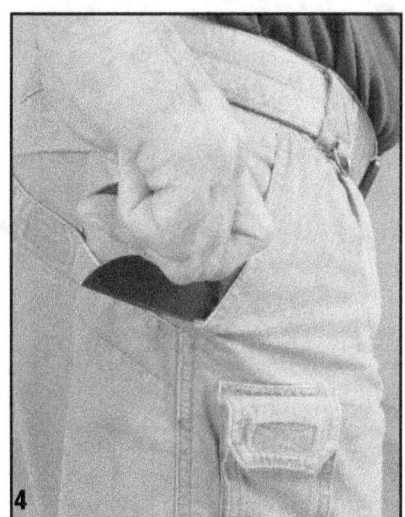

Step 3. Firmly grasp the folding knife

Step 4. Press firmly against outside edge and begin to withdraw knife from carry position keeping tension against outside edge of pocket material.

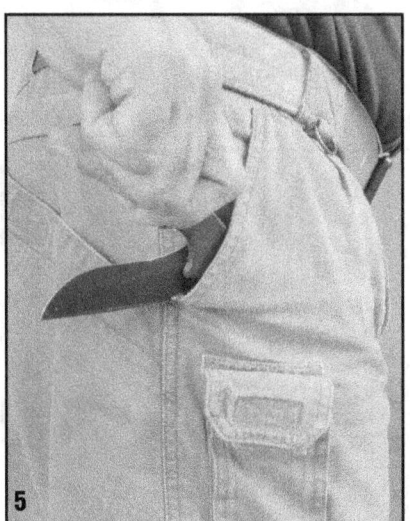 

Step 5. Pull up and back in such a manner as to engage the opening mechanism (wave). At this step the blade will begin to open.a

Step 6. Continue to withdraw the folding knife in order that the blade, using the friction of the wave (hooking of clothing fabric), locks into the open position.

Folding Knives: Carry & Deployment

Holes

Made popular by SpyderCo back in the 1980's, (akin to the invention of the carry clip) this particular opening mechanism was an innovation to the knife world. No moving parts, no additional gear, no requirements other than to place your thumb near or around the hole, give it a slight push and bam – blade on its way out from the handle and into an open and locked position.

Blade openings machined out of the blade can be just about any shape and size, however, you don't want to go too big, because when stored in your pocket keys and other pocket contents make get caught up in the hole and cause a series of unwanted events. A hole machined into the blade should also not be too small as your thumb may not fit and therefore fail to function as an opening mechanism.

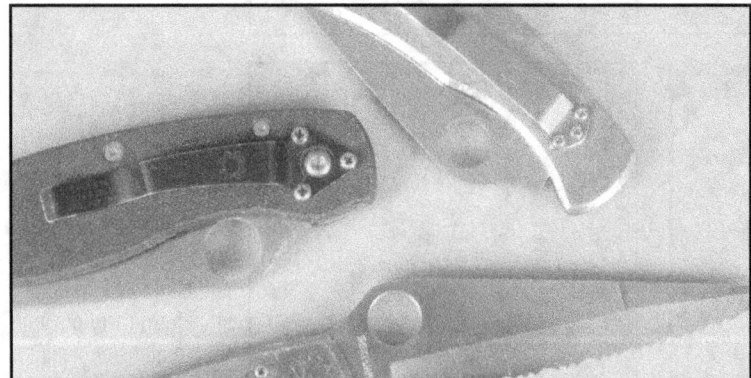

Samples of the innovative SpyderCo thumb hole

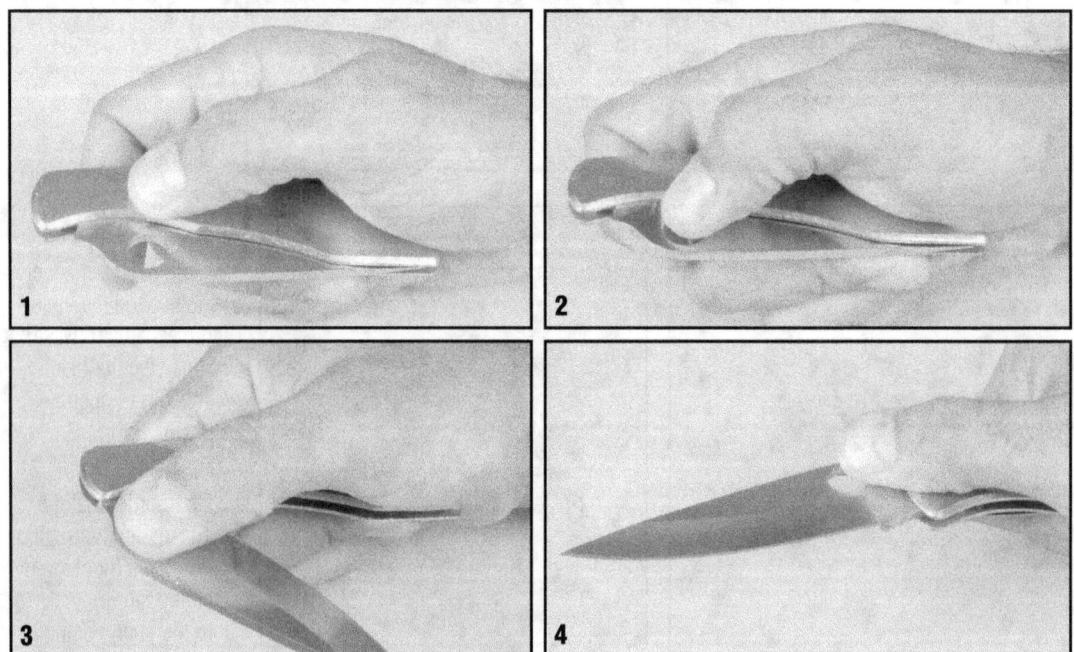

Example of utilizing a machined Hole opening mechanism.

Ovals

Same as the hole above, ovals machined out of the blade can be just about any shape and size, however, you don't want to go too big, because when stored in your pocket keys and other pocket contents make get caught up in the hole and cause a series of unwanted events. An oval machined into the blade should also not be too small as your thumb may not fit and therefore fail to function as an opening mechanism. Some users claim that the oval is a "better fit" for them, as covered earlier opening mechanisms are simply a matter of personal preference.

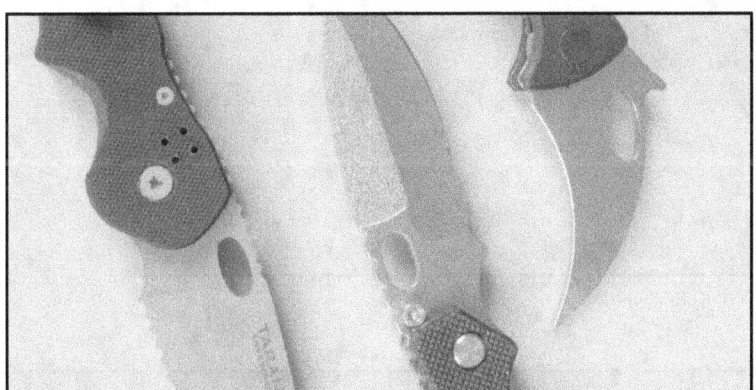

Samples of the Oval opening mechanism.

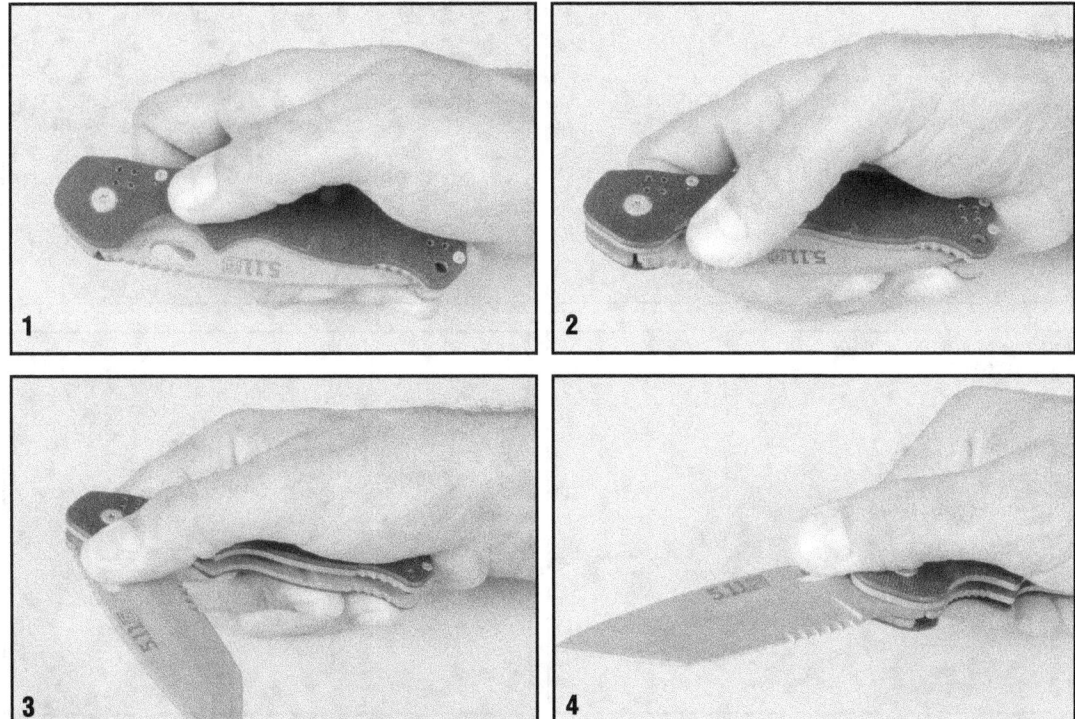

Example of utilizing a machined Oval opening mechanism.

Folding Knives: Carry & Deployment

Indents

Probably one of the oldest and most traditional opening mechanisms is the indent. Even back in ancient times the pocket knife was commonly opened by a simple indent struck into the "spine" or top (unsharpened) edge of the blade.

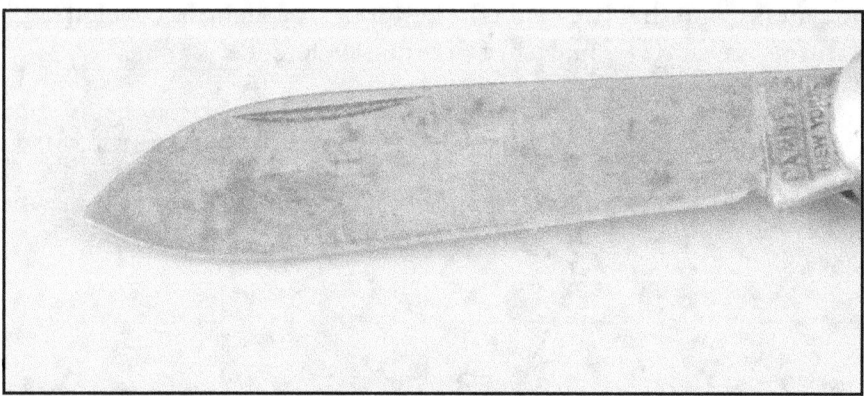

Samples of the Indent opening mechanism.

1	2
3	4

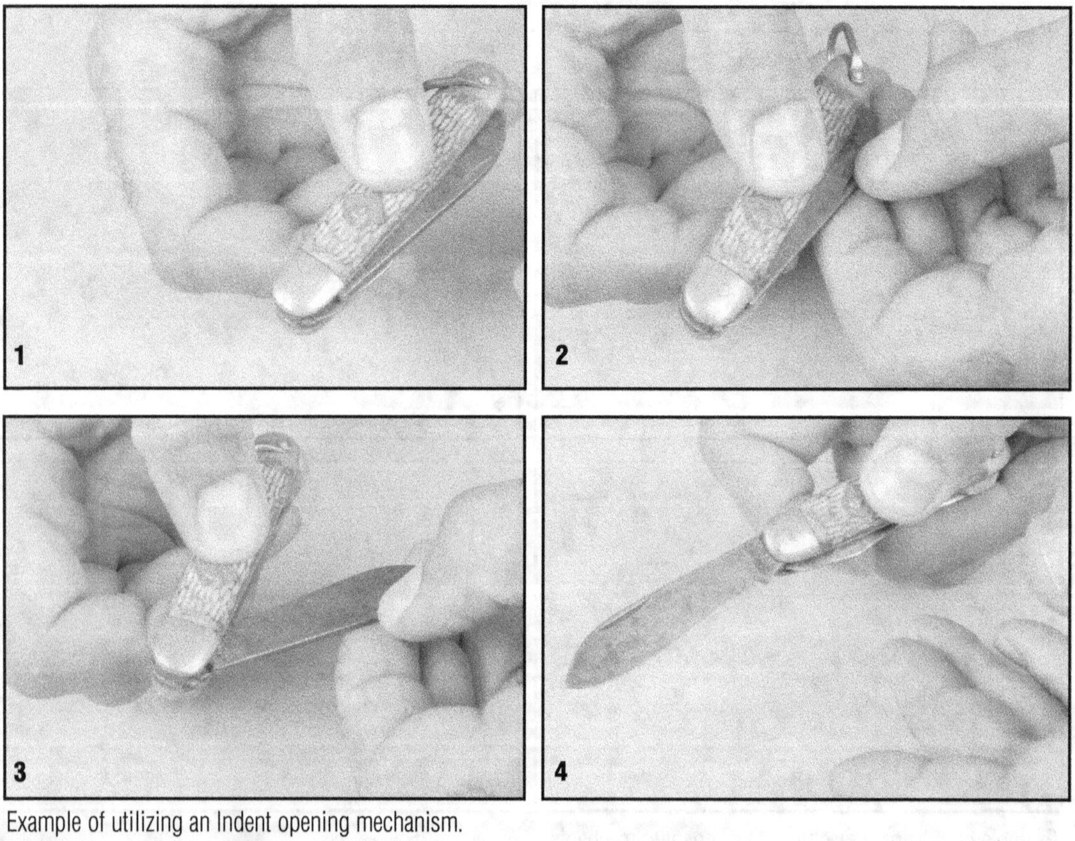

Example of utilizing an Indent opening mechanism.

Part II

Grooves

Similar to indents, grooves are among the oldest and most traditional (back in the days of Camillus and Buck Knives) of the opening mechanisms. Access of a simple groove etched into (or machined out of) the top of the closed knife blade was all it took to open the blade from the closed and unlocked position.

Example of Groove opening mechanism.

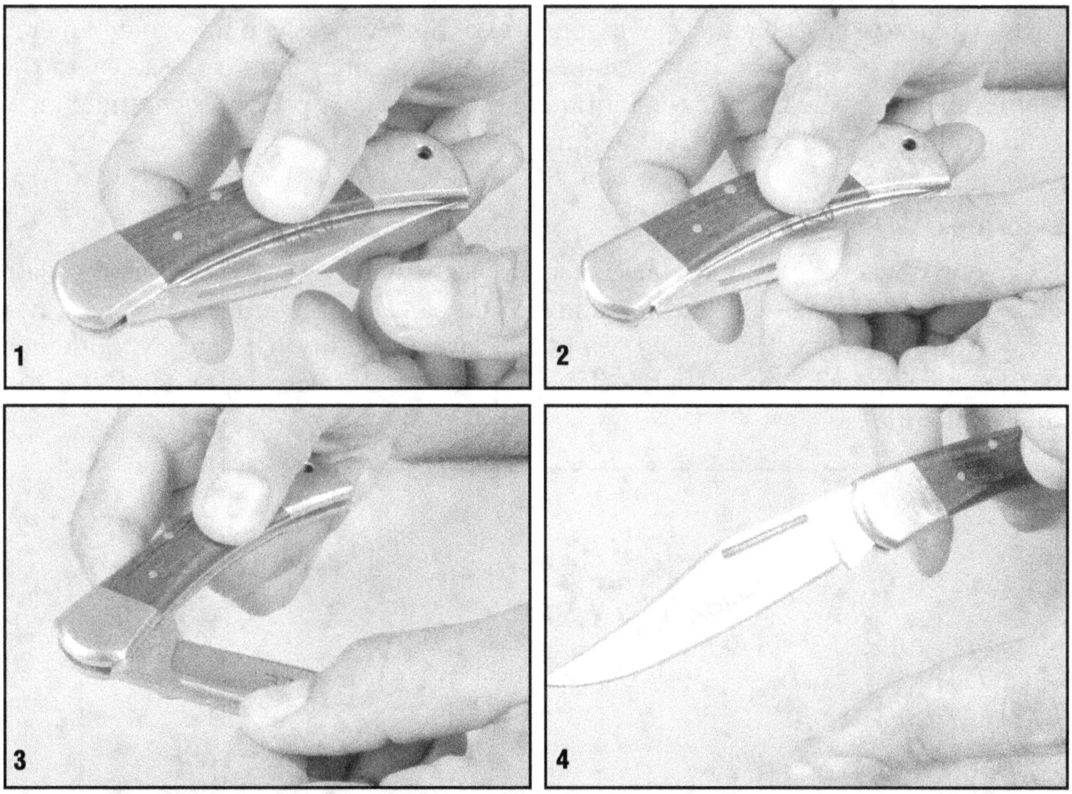

Example of utilizing a Groove opening mechanism.

Folding Knives: Carry & Deployment

Locking Mechanisms

Unlike the ancient Roman pocket knife (see Folding Knives - A Brief History - note that the pocket knife was invented prior to the pocket) and the 1890s Camillus among other earlier folders – there was no such thing as a locking mechanism (other than maybe a clasp or latch).

In fact it wasn't until the late 18th century before reliable cast spring steel would be produced. History tells us that spring steel was the invention of a Sheffield England clock maker named B.A. Huntsman in around the year 1742. Later on spring steel was utilized to produce back springs one of the earliest locking mechanisms (next step up from the very old school latch and clasp locks).

Today there are vast quantities of locking mechanisms available in the world of knives. Although there are unwieldy numbers, it suffices for purposes of our scope of study hear to list about a dozen or so of the more popular and mainstream.

Clasp Lock

The oldest known locking mechanism is the clasp. Simply a piece of material fastened to one side that clasped to a piece of material on the other side (from ancient text "similar to a brooch and cape"). Sometimes this was not even made out of metal. In ancient times this may have been leather and in some cases a precursor to the leather sheath. Some fragments of antiquity suggest the clasp lock was nothing more than a thin leather or cloth strap holding the blade closed.

Latch Lock

Later on down the timeline came the latch lock which was a step up from the clasp lock as it was a permanent fixture and part and parcel of the knife and at least had form-fitting parts which provided a minimal layer of additional security. Although not exactly bomb-proof, the latch lock has survived to this very day and in fact is still in wide use in Asia.

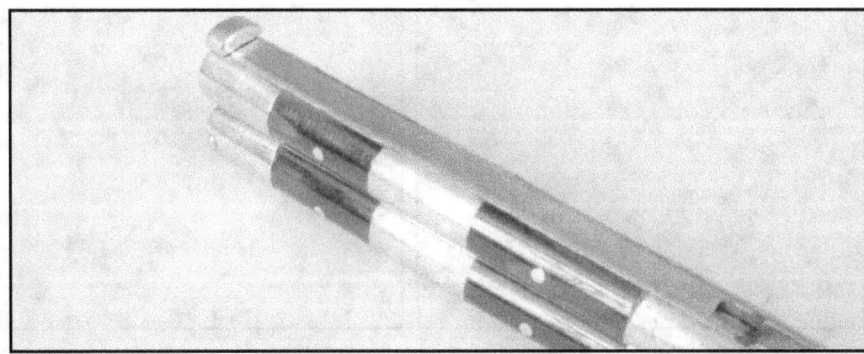

Example of the Latch Lock.

Part II

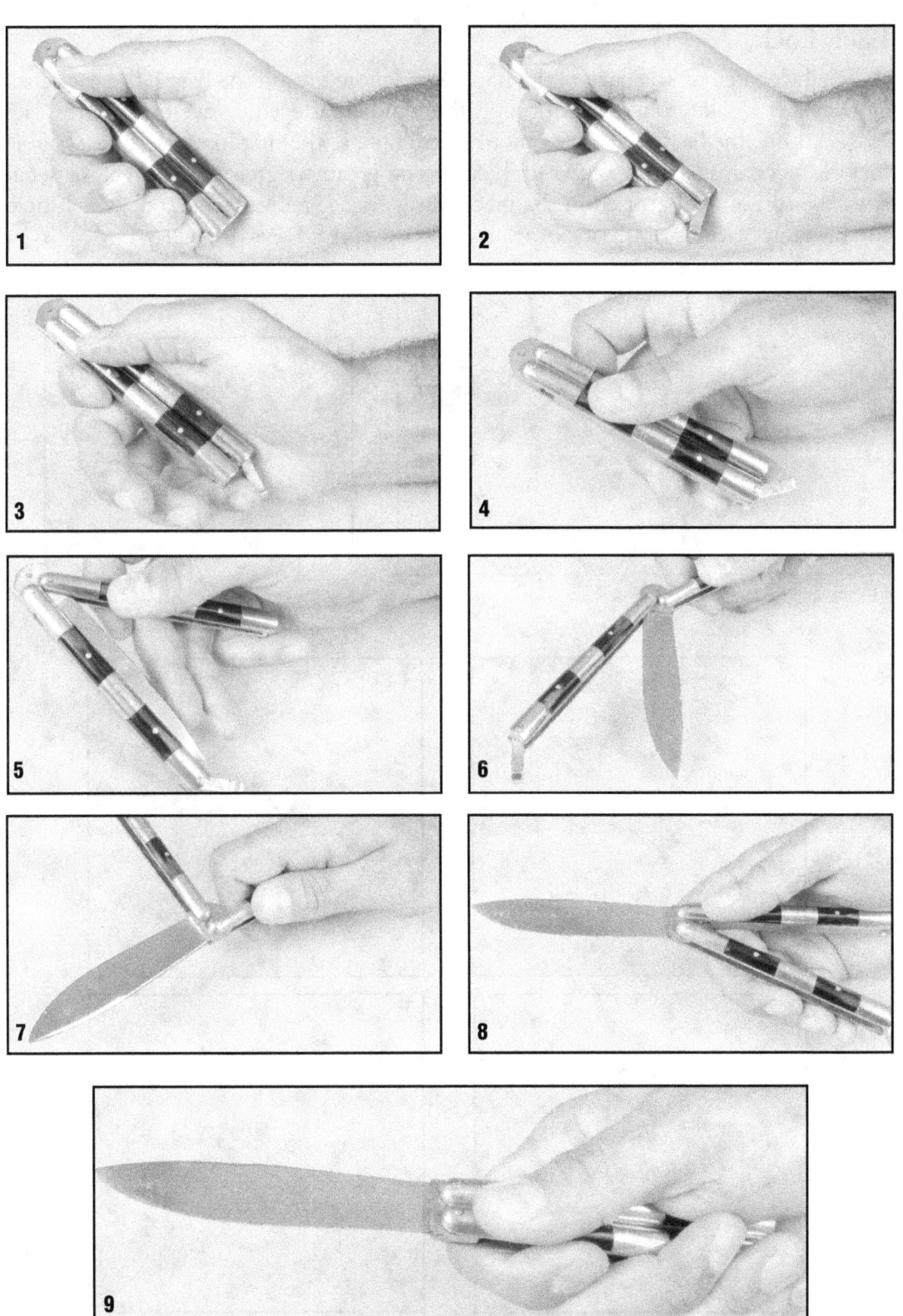

Example of utilizing the Latch Lock.

Folding Knives: Carry & Deployment

Back Lock

Usually located on the back of the handle and more toward the base – the back lock gains it's namesake from its position on the handle. Allegedly developed in the late 19th century, the back lock is commonly attributed as the first in a long line of stable locking mechanisms which allowed the user a far greater degree of blade locking security. The spring steel utilized in the mechanism engaged an internal lock which held the blade in place until the lock (again using the spring) was disengaged.

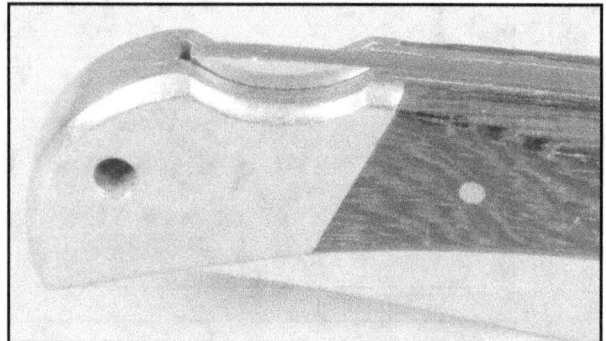

Example of the Back Lock.

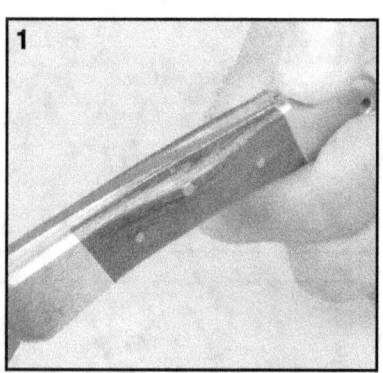

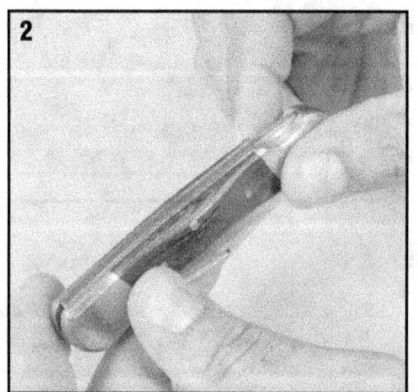

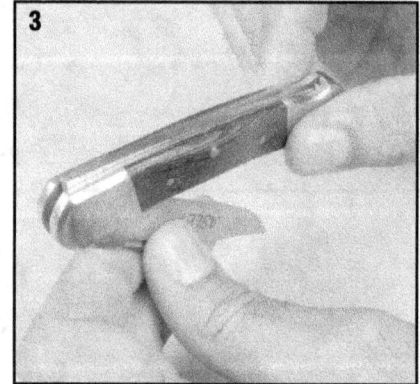

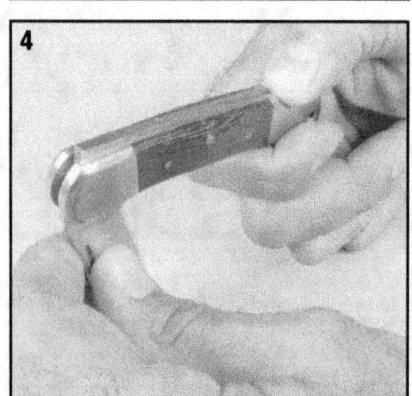

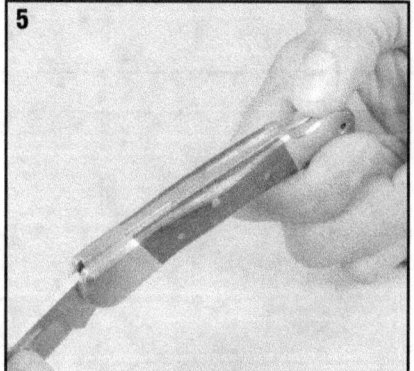

Example of utilizing the Back Lock.

Front Lock

Brain child of mechanical engineer Vince Ford, the modern front lock was allegedly developed by Blade Tech Industries in the late 1990's. The front lock (similar to the back lock in that it earned it's namesake by virtue of its physical location on the knife handle) included an extra internal steel bar that forced the lock into place and acted as a bolt across the spring firmly locking the blade in place at the base. This important added feature is critical to such folding knives as the curved blade. In fact many people complained when it first came out of how difficult it was to close the blade – not such a bad problem in my opinion especially when utilized with the curved blade!

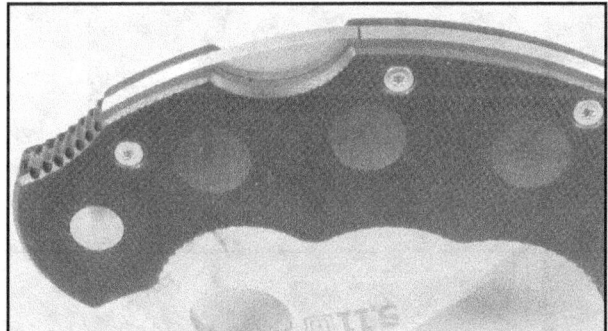

Example of the Front Lock

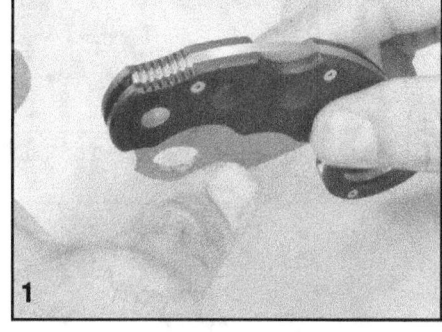

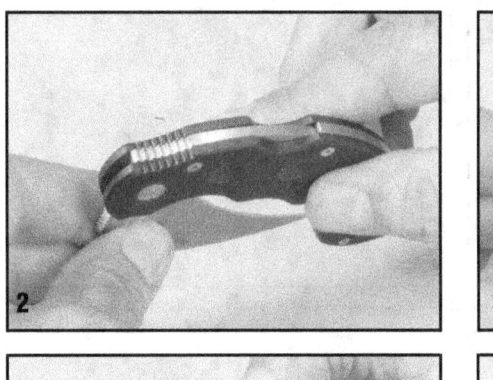

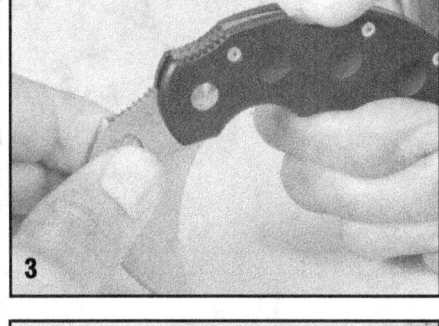

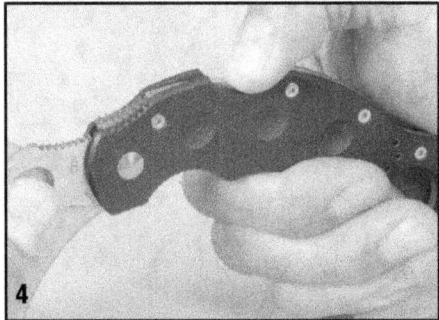

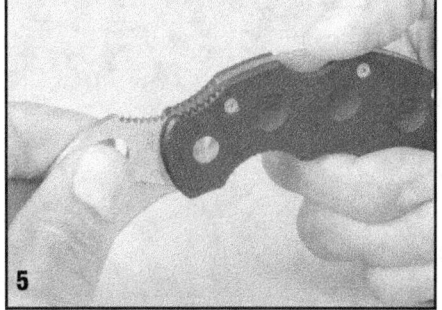

Example of utilizing the Front Lock.

Folding Knives: Carry & Deployment

Liner Lock

Internal to the handle the liner lock is exactly that – its part and parcel of the liner of the handle. In most cases the liner steel is strong enough to withstand a reasonable amount of force that may be applied to the blade (in either direction) to keep the base of the blade locked into place. It is disengaged by simply moving it back to its original "unlocked" position.

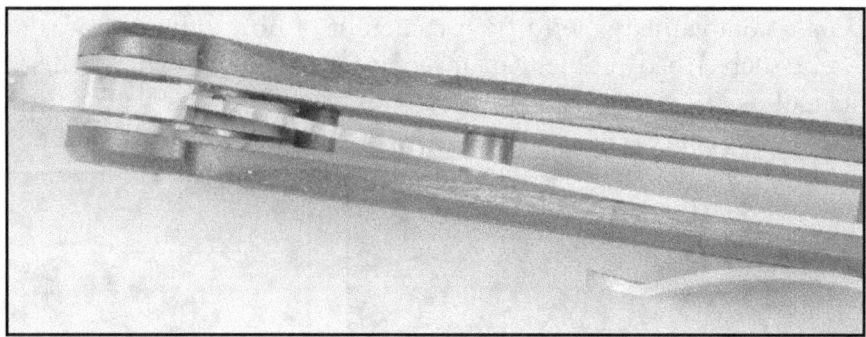

Liner Lock in the open and locked position.

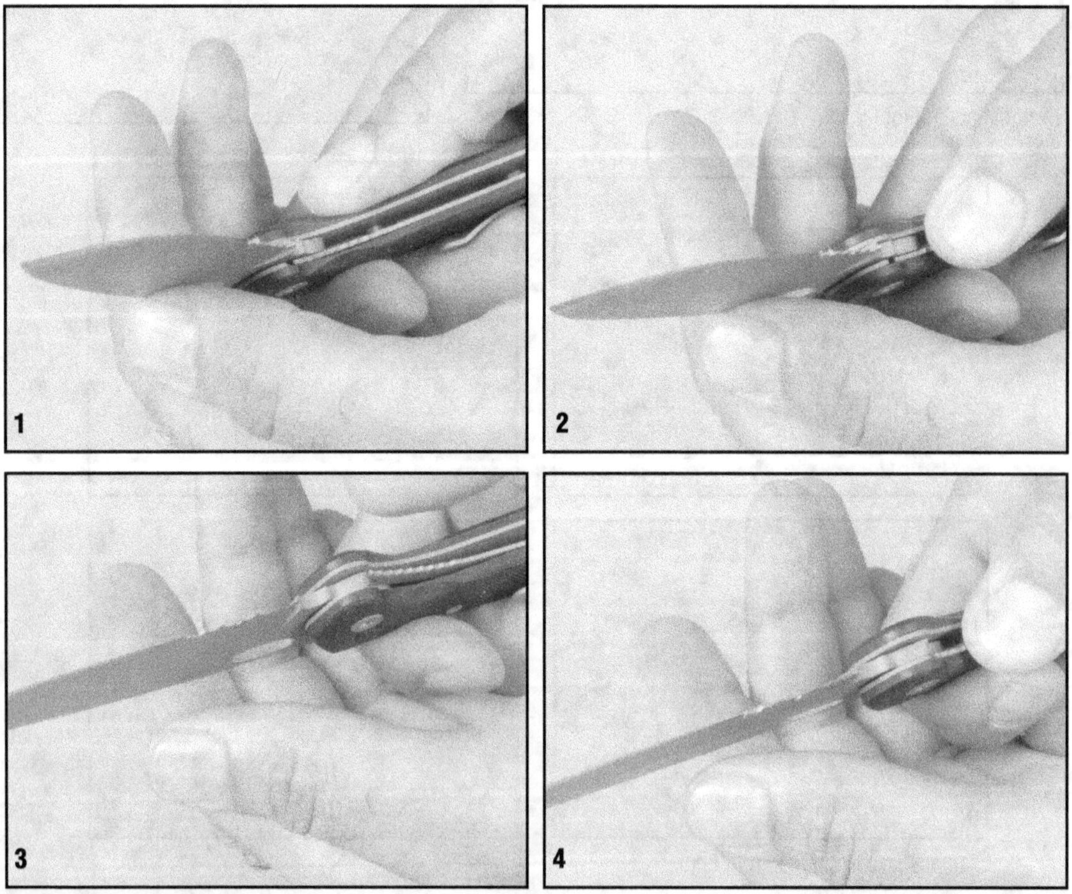

Example of utilizing the Liner Lock.

Part II

Frame Lock

Similar to the liner lock, the frame lock is part of the actual frame of the knife itself. Usually several times thicker than a liner lock, the frame lock is one of the sturdiest and most reliable locks out there with little or no moving parts. One of the key advantages of the frame lock is the fact that when it is engaged the lock is additionally supported by the user's hand which is held firmly in place by the actual grip of the user.

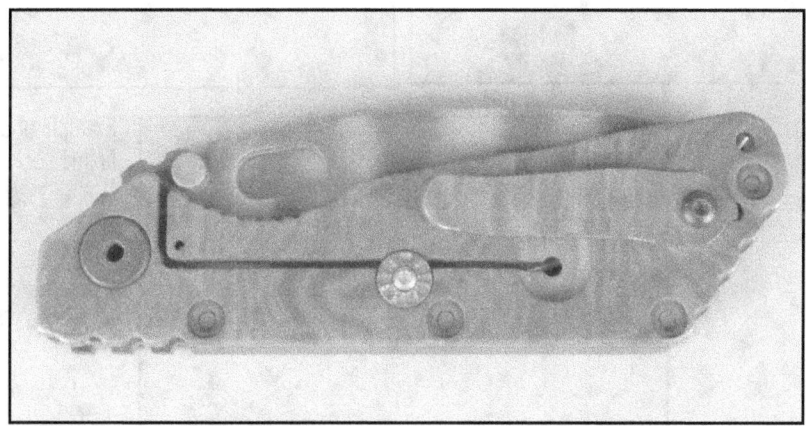

Example of a quality Frame Lock.

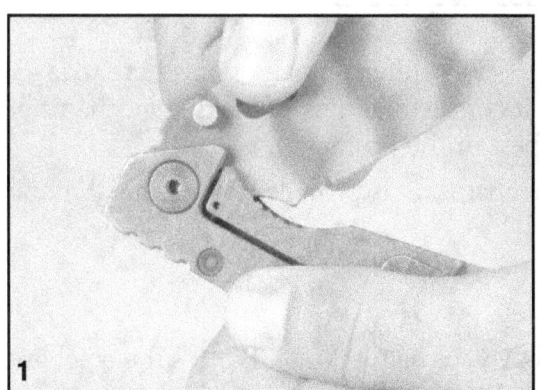

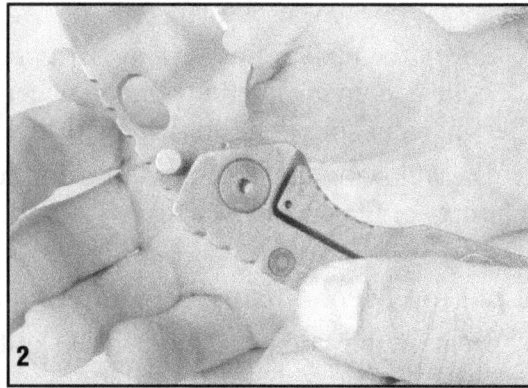

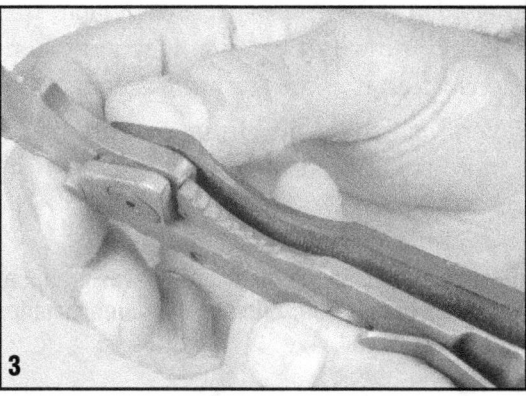

Example of engaging the Frame Lock.

Folding Knives: Carry & Deployment

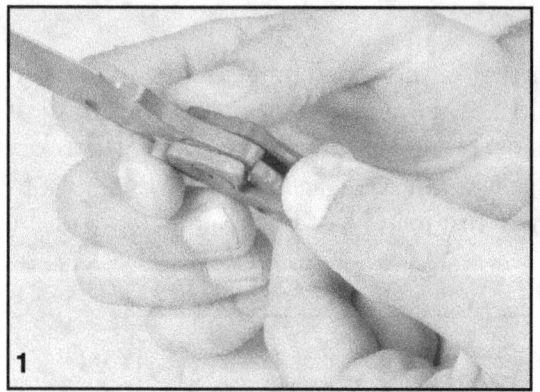
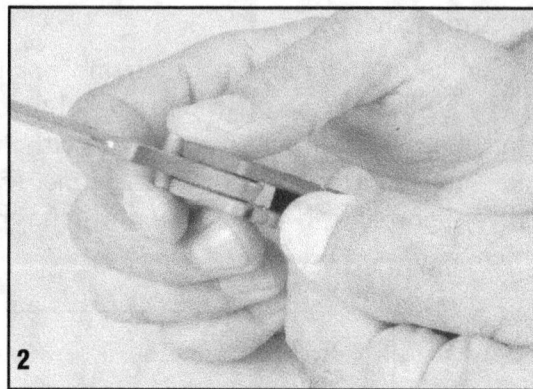
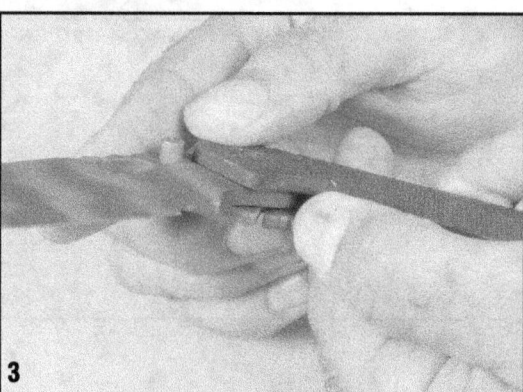

Example of disengaging the Frame Lock.

One of the more innovative modern production knife companies out there today is Benchmade. In my humble opinion, when it comes to variety of locking mechanisms Benchmade ranks right up there with the best in the business.

The following list of modern locking mechanisms is provided by permission and courtesy of Benchmade® Knives.

Locking Mechanisms By Benchmade®

Seems simple enough. You open the knife, you close the knife. But it's the mechanics of motion that happen in the process, which can make the difference between ordinary and extraordinary. Is the knife easy to actuate, can the knife be opened with one hand, is it ambidextrous, and will it be reliable when you need it most? These are all essential considerations when it comes to knife mechanics.

Benchmade offers several mechanisms, from patented exclusives including the unparalleled AXIS, to more traditional lock back designs. It all comes down to your level of desired function.

Note: Following text and illustration of Locking Mechanisms was produced by Benchmade Knives and included in this manuscript by permission. Click on www.benchmade.com for more info. © 1987-2007 Benchmade. All rights reserved.

Part II

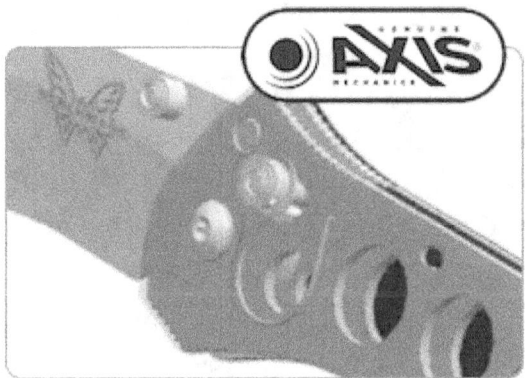 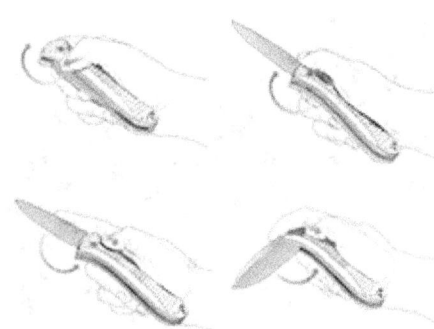

AXIS® A patented Benchmade exclusive, AXIS has been turning heads and winning fans ever since its introduction. A 100-percent ambidextrous design, AXIS gets its function from a small, hardened steel bar which rides forward and back in a slot machined into both steel liners. The bar extends to both sides of the knife, spanning the liners and positioned over the rear of the blade. It engages a ramped, tang portion of the knife blade when it is opened. Two omega style springs, one on each liner, give the locking bar it's inertia to engage the knife tang, and as a result the tang is wedged solidly between a sizable stop pin and the AXIS bar itself.

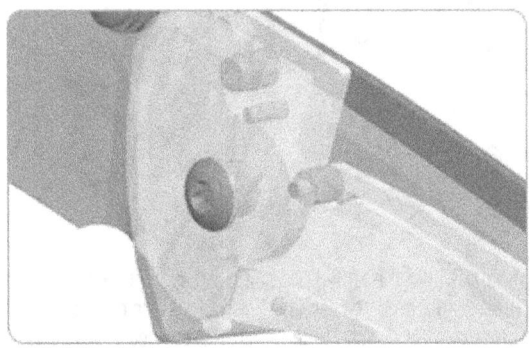 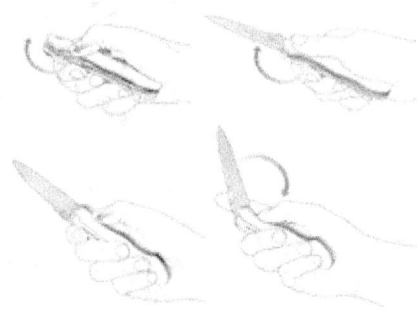

NITROUS™ Another patented Benchmade exclusive, the Nitrous boosts the blade with ease. As the blade is closed the two torsion arms which run the length of the handle liners are secured in place and make contact with the blade tang. As tensioned against the blade tang, the user rotates the blade open to a 30-degree angle, the torsion arms take over and continue the blade opening process on its own. The huge advantages to the Nitrous design over other similar concepts, is that the blade must be rotated open beyond a 30-degree angle, which offers added user control.

Folding Knives: Carry & Deployment

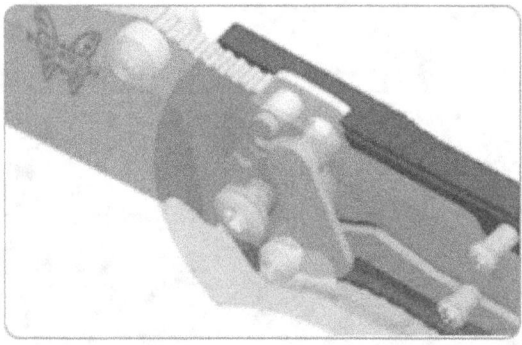 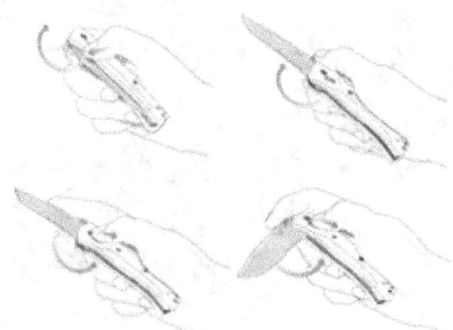

ROLLING LOCK® WITH INDRAFT® Another patented Benchmade exclusive, the Rolling Lock utilizes a sizable, notched lock-pin, which secures against the blade tang when engaged. To disengage, a spring actuated thumb button on the right handle side is drawn back to rotate the lock pin and free the blade. InDraft is a patented exclusive, which is a combination of a slotted liner and a pin in the blade tang working in tandem to generate an inward inertia when closing the blade. This results in one of the safest blade detents available today.

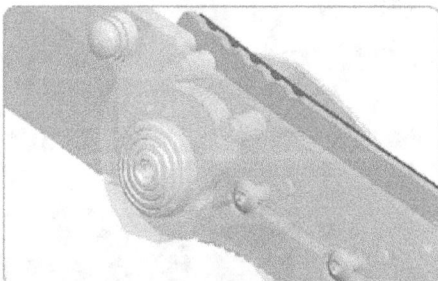 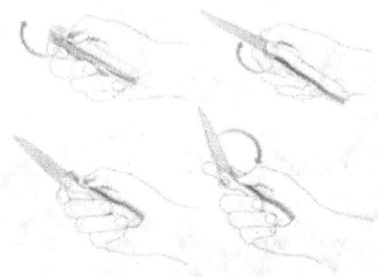

MODIFIED LOCKING-LINER The basic principle is an integral locking bar within the knife liner being stress bent, enabling it to spring into position behind the rear tang of the blade when the blade is opened. The locking bar wedges against the rear of the blade, locking it open until you physically push it clear and close the blade. With little practice, the process of opening and closing the blade can be done single-handed (locking-liners are made either right or left hand specific). And with a Benchmade modified locking-liner you get a patented feature, which helps to enhance the lock function. The function success and function failure of a locking liner depends greatly on how well it is made and also the quality of the materials used. With Benchmade, you get the best of both, and we offer it in a variety of knives.

Part II

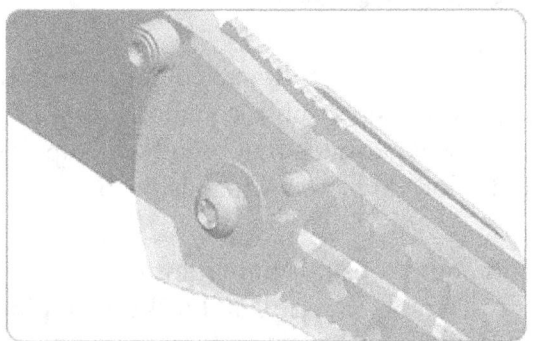
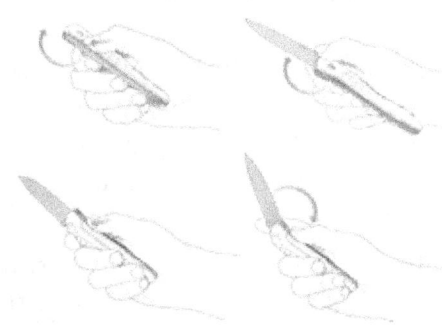

MONOLOCK The monolock mechanism is basically a locking-liner on steroids. The knife liner is one in the same as the knife handle, and thus it is designed and made to function as the locking mechanism. Subsequently, a thicker material is used to provide enough surface area to be a functional handle, and in turn creates a larger surface area to lock the blade with. If executed properly, the monolock rates very highly in strength and function.

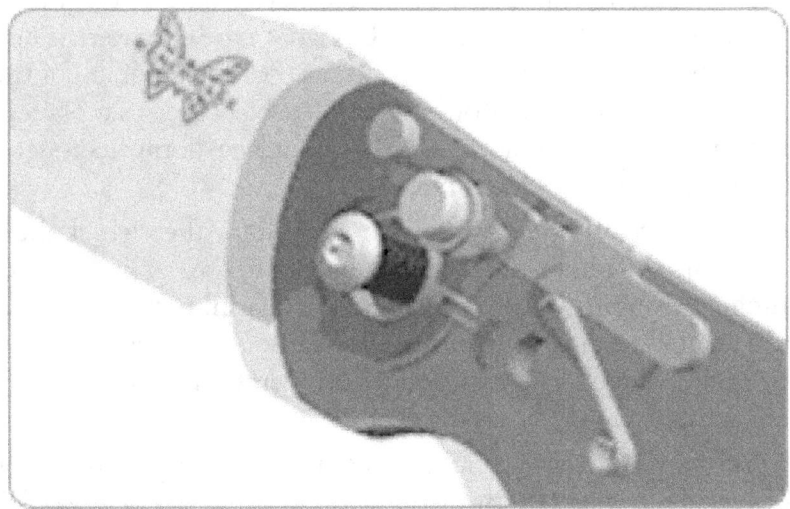

AUTOMATIC It's important to first note that these types of knives are restricted to Active Duty Military and Law Enforcement, and Federally regulated. Be certain of permissibility in your location including state or local governments and within your reporting bodies. These are folding knives, which open at the push of a button... An especially important feature to have if your disability, occupation or profession has the potential of limiting your mobility or freedom to access a knife through traditional means. The blade is deployed out from the handle side like traditional folding knives, but the blade is spring open assisted to a fully locked position. Benchmade is recognized for making some of the finest auto opening knives in the world. These are primarily manufactured for Law Enforcement, Military and Government Agency distribution. The Benchmade autos take full advantage of a powerful coil spring blade actuator coupled with our precisely machined taper lock mechanism, proving itself to be the most reliable system available in auto open knives. And our patented integrated safety mechanism (located on handle spine) ensures that the blade will not inadvertently deploy or close once it is locked open. Benchmade autos are precision engineered and manufactured to perform in the real world, you would be hard pressed to find a superior tool.

Folding Knife Selection

As a full-time professional instructor throughout the US military, federal, state and local law enforcement training communities, the single most common questions regarding folding knives is selection. The same two questions that usually pop up are "What is the best folding knife out there?" and "How do I know which is the best folding knife for me?"

In order to answer these two critical selection questions it is important to take into account several considerations with regards to owning, carrying and utilizing a folding knife. These considerations include, but are not limited to: functionality and application, quality and cost, size and the law, accessibility, appearance, policy, aesthetics and nomenclature, ergo dynamics and personal preference. Let's take a closer look at each of these important considerations.

First and foremost in the selection of a folding knife is functionality. What is the knife going to be used for? Is it a work knife or a recreational knife? Will it be used for both work and recreation or one and not the other? And if it is a work knife, then what exactly will it be used for on the job? On what type of materials do you plan to use it? Rope, seatbelts, rubber, nylon, cotton, cardboard, leather, all of the above? As we have already studied, the question of a serrated blade versus a plain edge is additionally a consideration of functionality. The old adage "form fits function" directly applies to the selection of a folding knife.

The questions of overall application such as: What's the right knife for my job? What exactly is this knife to be used for? Am I going to use my knife as a utility blade? Am I planning on using it for self-defense? Am I planning to use it as a work tool? Maybe even a combination of utility and self-defense? "What is the primary function of my folding knife?" should be a paramount consideration to determine the selection of make, model and style.

As mentioned above (see Model Policy) quality is also an important consideration. In some cases maybe to practice throwing a knife into a tree you would most likely not want to use a $2,700 custom folding knife to gain these skills. Maybe something of extreme low quality as you know the knife will eventually be destroyed by its intended use. On the other hand, if part of your job responsibilities call for the usage of a knife on duty, then there can be no compromise on the level of quality – except with regards to affordability.

While we're on the topic of affordability, cost is the bottom line for most of us. In selecting a folding knife there is that fine line of high quality versus cost. We generally seek the "best bang for the buck" which is a healthy approach to selecting a folding knife. However, those of us with an addiction to fine cutlery may find ourselves with more than one knife that we want as opposed to need! In most cases there is a direct correlation between cost and quality.

Generally speaking the more expensive the knife, the higher the quality - the old adage "you get what you pay for" in most cases does in fact does apply to the world of modern folding knives. For example a custom built, hand-ground, American-made S30V CPM steel blade that has been cryogenically heat-treated to an above-standard Rockwell rating will cost substantially more than a stamped out piece of 734 steel hailing from a mass-production knockoff sweat-shop in mainland China. The bottom line is of course your personal budget. What level of quality will that budget allow?

Is the possession and carry of folding knives permissible by law? If you lived in England you would be breaking the law carrying a folding knife in public. When traveling abroad it's important to check with local authorities regarding the possession and carry of a folding knife.

With regards to the law, one of the most basic considerations is size. What is the optimal size of the blade, the handle and the overall knife from blade tip to the base of the handle? When considering blade length the first and foremost concern should be compliance with the law. What are the laws in your state, city or municipality with regards to the length of a folding knife, predominantly the blade and in some cases the overall knife length including the blade and handle.

Historically, it was documented by the Western European duelists of the 16th century that "the length of a blade at only four inches is a most sufficient length [in the hands of an expert] for one man to take the life of another man." From that time on, referencing a knife blade the length of about four inches was pretty much generally accepted by law as a "substantial length."

Centuries later with the establishment of the colonies in America and the subsequent forming of the states of the Union, there were virtually no changes to this notion. In fact even to this very day a number of States require that the length of a folding knife be "no longer than three and one-half inches." The old-world quote of the duelist is the reason for this decision.

As laws regarding edged tools sometimes change like the direction of the wind, it remains incumbent upon the owner of the knife to stay abreast of the latest changes. In our modern computer age and in addition to checking with your local police department and library regarding the law, there is also the option of the internet. The following are a few sites available online at the time of this writing that apparently provide useful information with regards to the law and the length of a folding knife blade.

http://www.thehighroad.org/library/blades/knifelaws.html
http://www.knifelawsonline.com/knifehome/
http://www.akti.org/legisl.html
http://www.cutleryscience.com/reviews/laws.html

Folding Knives: Carry & Deployment

Next on our checklist is accessibility – how accessible must your folding knife be (readily accessible, immediately accessible, not very accessible, etc.) and given the answer to that question, how and where can it be carried. On the question of accessibility, this would be determined by the application of the knife. If you are a first responder and your job includes cutting shoelaces and seatbelts from victims of car accidents and other related emergency cutting, then accessibility to the knife would be paramount. The folding knife would need to be of solid construction with a heavy duty anchored carry clip which allowed immediate access under stressful conditions. Later on in our study we take a much closer look into the area of folding knife carry, in particular how and where to carry.

Depending upon your job and work environment as well as off duty and in a social setting, appearance of a folding knife is also a consideration. If your job requires formal business attire and co-workers and clients observe a large unfamiliar business accoutrement clipped next to your cell phone on your _" belt in the shape of a huge oversized folding knife anchored with a one-inch thick steel carry clip, it may lead to adverse opinion or in some extreme cases be a detriment to your career. In that case it may be a better choice to go with a more streamlined folding knife design which better fits business attire.

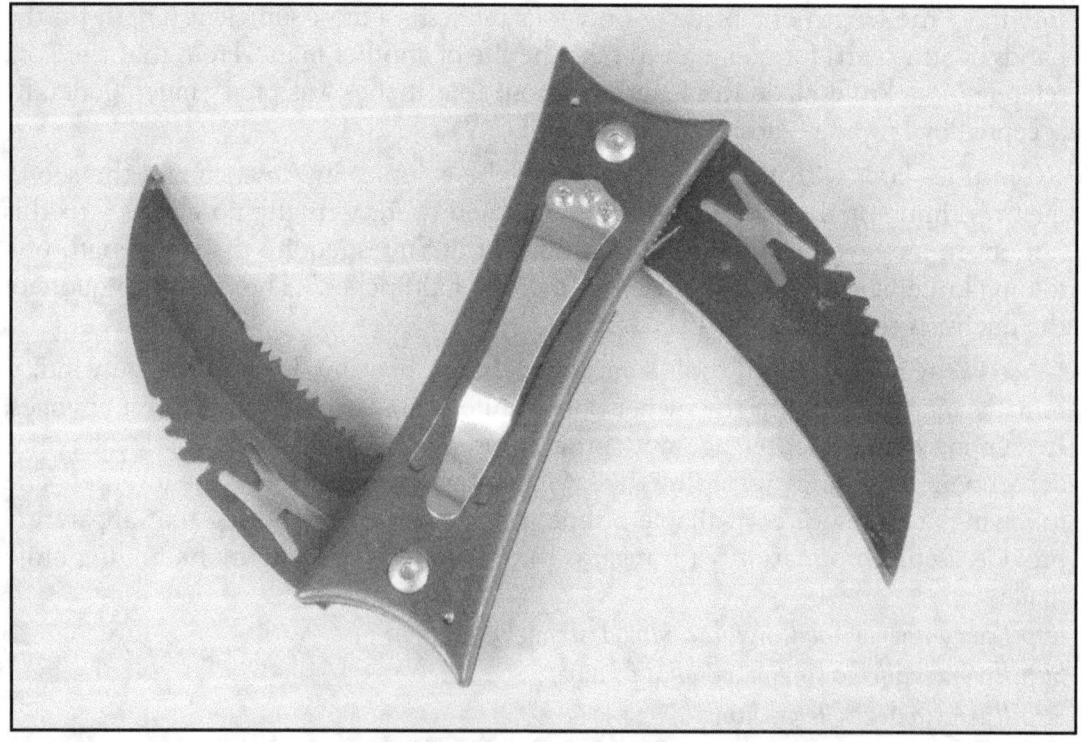

A $7 folding knife purchased from a gas station on a very busy street corner in downtown Detroit, Michigan.

Part II

In the world of law enforcement (federal, state, municipal, etc.), agency or department policy plays a critical role in the determination of gear selection. As an employee you may be issued a knife – extremely rare, but in most cases as was covered above (see Model Policy) it is incumbent upon the individual officer or agent to decide on a particular folding knife provided that it falls within departmental policy and equipment guidelines.

As per above, it must be kept in mind that most departments select NOT to maintain a policy with regards to folding knives as this deflects the onus of responsibility away from the department and landing it squarely on the shoulders of their employees. In such cases it's a good idea as an employee carrying a folding knife on the job to run it by supervisory personnel and at least have some idea of what the department would consider "acceptable" in lieu of a formal knife policy. Once this determination has been made it at least can provide some rough guideline to assist you in your folding knife selection process.

If a folding knife is not carried concealed, then it is readily observed by the general public. If a knife remains in plain view of the general public, then aesthetics is a real consideration. Imagine that you are a patrol officer and come in contact with the public on a constant and regular basis and your folding knife has a skull and crossbones emblazoned on the handle sticking out of your uniform – it may in some cases illicit certain uneasiness with those who notice it. Plus, how would your supervisor react to an inadvertent photograph of it that may end up in the newspaper and what would the chief say about it? In much the same vein, nomenclature plays a significant role in the folding knife selection process even in the private sector. In the event that your knife has the words "Ninja Death Blade" scribbled out in blood-red lettering against a dark black background, how would your co-workers react and what would your clients, customers and even other family members say about it? These are important points to keep in mind when shopping for the optimal folding knife.

Nomenclature with regards to department interests should also be included as a consideration. In the world of knives there are a number of terms which can be applied to folding blades which may be used for police work, it is important for administrative documentation to refrain from such fearsome nomenclature or terminology as "weapon," "bladed weapon," "edged weapon," "combat folder," "cutting blade," "tactical blade," etc. for obvious reasons. More acceptable terminology would be: "Rescue Knife," "Police Knife," "Utility Knife," or "Duty Knife," which are synonymous and may be used interchangeably with respect to police work. It is important to note that certain manufacturers may label their products inappropriately with regards to department interests such as "Combat Folder," "The Defender," "Fighting Knife," etc. More suitable are such product descriptions as "Police Knife," "Police Advocate," "Rescue Knife," "Rescue Tool," "First Responder," etc.

Ergo dynamics (comfort and fit to the body): One of the last remaining questions remaining is truly subjective: Does it really fit my hand? Does it feel right? Is it comfortable to grip, easy to open and close? Is the weight and width right? Does it feel like good ergo dynamics? These are all truly subjective questions and are subject to personal choices of the buyer.

Last, but certainly not least is the consideration of personal preference. Selection of a folding knife is as personal as that of clothing. Yes such considerations of functionality and application, quality and cost, size, the law, accessibility, appearance and aesthetics can certainly assist in making an appropriate decision, but you're the one who carries and uses the knife every day. You're the one who needs to be happy with it. Your boss, co-workers, customers and peers may all have their own opinions, but if you're within the boundaries of the law and "form fitting function" and given all of the other considerations listed above it does in fact boil down to the simple determination of personal preference.

Preventative Maintenance (PM)

All things man-made break. This is simply a law of physics. Knives are no exception to this universal law. As a general rule it is far better to be proactive than reactive to potential operational problems. Preventative Maintenance (often abbreviated PM) refers to performance of such a proactive maintenance in such a manner as to prevent operational problems.

A knife isn't exactly a complex piece of mechanical gear, but similar to not changing the oil in your vehicle and never ever cleaning the utensils from which you eat – lack of PM may possibly result in a lack of functionality. If your job requires the reliable usage of your folding knife, then this certainly becomes a concern.

PM is one of the most ignored aspects of firearm ownership so you can imagine how low on the priority scale PM on a knife may fall. PM saves money and time, it certainly improves performance and especially if you may be a First Responder, may even save someone's life.

Sharpening

One of the most important aspects of preventative maintenance is sharpening. Although most modern production folders these days can literally go for months and even years without sharpening, what is the use of a knife if it's dull?

Sharpening is literally an art form and to do it like the professionals it would take years to develop those skills. But, for the majority of us who own modern production folding knives there are a number of sharpening devices available today from which to choose.

The classic handheld sharpening stone is readily available, a low-tech, low cost solution and is easy to use. On the opposite end of the spectrum there is the professional knife grinder's wheel. By way of handheld rods there are sharpening steels and sharpening ceramics which are also low-cost, but really work and are easy to use. There are simple-to-use kitchen electrical sharpening machines, these also work on folding knives. Such sharpening tool companies as Lansky, Diamond, SpyderCo (which is also a production folding knife company), Wusthof, J.A. Henckels, Brieto, Zwilling, Gatco, Buck Knives (one of the oldest knife companies in America), Global, and many others. If you have access to the internet you can spend all day running through literally hundreds of sharpening devices.

Bottom line on sharpening – what I do is send it back to the manufacturer and usually they will sharpen it for you and most of the time at no cost – provided you pay for shipping and handling. Other than that, a simple low-cost, low-tech, easy-to-use solution is the best as I'm a devoted proponent of "keep it simple."

Lubrication

Very similar to a gun, it's important to keep your edged tools cleaned and lubricated. After many years without proper lubrication the bushings at the pivot screw can wear down to the point of useless. Just a drop of oil every once in a while will last for a very long time.

Cleaning the blade is often ignored. With serrated or combo edge you really don't want to collect any gunk and grime as it will significantly impact the performance of the cutting edge. Wiping the blade down with a slightly oiled rag and then wiping it completely clean after that will usually keep the blade protected and performing well into the future.

Storage

If you're not planning on using your folding knife for a long time, then it's a good idea run a little light knife oil on both sides of the blade and a drop on the pivot screw (no greases or anything too heavy as it may gunk up over time). This will assist especially in humid areas or near the ocean.

Another important tip is to not store your folding knife in any type of soft plastic or leather in humid or extremely volatile temperatures. Over the years, various handle materials can react chemically with various wrappers and animal byproducts and leave you with something that you'd need to send out to the lab for analysis.

Part III
Operational Proficiency

Part III

Operational Skills

The most important reason for owning a folding knife is obviously to use it. Part III of this manuscript is devoted entirely to the development of proficiency in the usage of the folding knife. Since it is the case that this manuscript is geared mostly toward the professional training community (military, federal, state, etc.) and predominantly law enforcement application (with regards to model policy), all usage of the folding knives for any other purpose than as a utility tool has been intentionally omitted.

If you happen to not be in the military, or a peace officer or federal agent, then you're also able to tremendously benefit from this same body of knowledge and information. Other than a few mission-specific items which have been omitted for various considerations, the majority of this data is what is taught at most mainstream academies and specialized advanced officer training programs.

If you are sworn and are concerned about the information here going out to the bad guys as well as global terrorist organizations – no worries, this material has been publicly scrubbed so very clean that it squeaks when you read it. Additionally, the material herein coincides with industry standard training requirements with terminal learning objectives for each training block equally matching these requirements. In other words, (loosely translated for us knuckle-dragging alpha males), all this stuff matches policy, works for real and is easy to learn.

Training Methodology

One of the most effective training methodologies out there is a copyrighted business methodology which is reprinted here by permission of the Operational Skills Group, LLC of Seaside, California (for more information click on www.opskillsgroup.com).

The instructors at the Op Skills Group (OSG) are basically one giant mobile training team that has been providing training services to federal, state and local agencies since the mid 1990's. Literally thousands of military, law enforcement and federal agents have gone through hundreds of professional courses delivered by OSG staff. It was found by trial and error over the many years and many students, that there is an optimal instructional method for delivery of physical skills training materials. In this following section, that methodology will be utilized to deliver these same materials.

Motor Skills

In any physical skills there are only two types of motor skills – Fine Motor Skills (FMS) and Gross Motor Skills (GMS). FMS are basically any skill that requires fine dexterity or manipulation of the fingers and hands in an intricate and precise manner such as threading a needle or tying your shoe laces.

Folding Knives: Carry & Deployment

GMS are any skills which require gross movement usually larger muscles and full limb movement. Such activities as clapping your hands, raising your arm over your head or moving to a squatting position would be considered gross motor movements.

According to the experts in the fields of kinesthesiology (the study of sensory motions mediated by "end organs" located in muscles, tendons and joints and stimulated by bodily movements and tensions) and physiology (a branch of biology that deals with the functions and activities of life or of living matter such as organs, tissues, or cells, and of the physical and chemical phenomena involved), under duress, the blood in our bodies tends to move from the delicate extremities (hands and feet) accumulating around the larger muscle groups.

The reflexive process of blood pooling around the larger areas and moving away from the smaller ones is one of the reasons most physical training focuses primarily upon GMS as opposed to FMS. In this next section the same reasoning applies. Emphasis has been placed on the execution of these maneuvers in such a manner as to focus predominantly upon gross motor skills as opposed to fine motor skills. Please keep in mind that a first responder may need to use a folding knife in a rapid and controlled manner (for example suicide prevention – cutting somebody down from hanging themselves with a belt around their neck) and in most cases must perform their job under physical duress – elevated heart rate, screaming babies, sirens blaring and radios sending and receiving traffic.

Folding Knife Carry

Prior to the era of the "combat folder" all folding knives were considered "pocket" knives. The reason for this consideration was that there were only two places you could carry your folding knife and that was either in a pouch strapped usually to your belt or in your pocket – there were no other options for folding knife carry.

In fact it was a family-owned knife company started by a Kansas blacksmith apprentice named Hoyt Buck way back in the early 1900's eventually called Buck Knives – perhaps one of the oldest knife companies in America today that came up with the concept of improving the "pocket" knife.

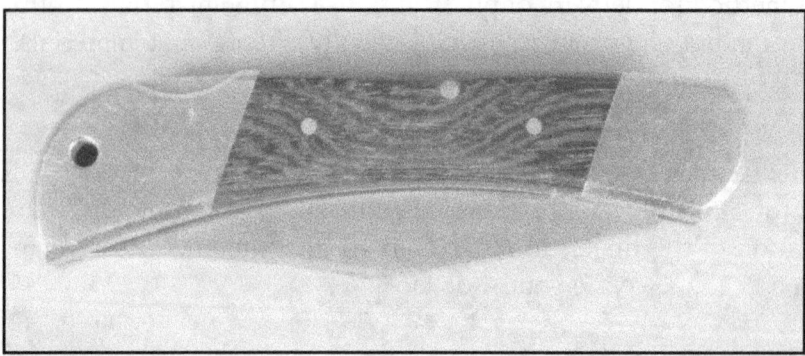

Typical folding knife configuration prior to the invention of the carry clip.

Part III

According to the company archives Al Buck designed the "Buck Model 110 Folding Hunter" in 1963. This particular pocket knife was extremely popular back in the day and came with a great leather sheath that fit the knife perfectly (those of us old enough to remember probably still have that knife and sheath to this day). The Buck Model 110 would also fit in your pocket nicely if you didn't have a sheath for it.

Roughly a quarter of a century (plus or minus) went by before the next innovation in pocket knives which was the addition of the carry clip. The carry clip, a form-fitted piece of thin spring steel fixed to one of the outsides of the handle, allowed the knife owner to carry the folding knife not only in the pocket but now actually on the pocket. Why was this such an innovative idea? It allowed greater accessibility to the folding knife.

Accessibility

As necessity is the mother of all inventions, the invention of the folding knife carry clip made obsolete the need for fishing around in your pocket for your knife with your entire hand hoping to grab it in the right place to try and grip and open it in a timely manner. Gone were the days of rooting around in your pockets or trying to unsnap your leather sheath to access your folding knife.

A typical carry clip can be manufactured out of various steels. Any of the quality clips available will of course be made of hardened steel.

By the way, there's absolutely nothing wrong whatsoever with pocket knives minus a clip. Some of the finest pocket knives I've ever seen are still made to this very day without a carry clip and are often used literally as a "gentleman's pocket knife."

Examples of modern Carry Clips

Folding Knives: Carry & Deployment

There are a number of collector's knives as well of outstanding workmanship and breathtakingly high quality – all without the carry clip. However, for purposes of the scope of study for this manuscript we will limit our dissertation to the carry and usage of the modern production folding knife complete with carry clip.

Viable Locations

Carry locations can vary considerably based on operational requirements. In the case of professional first responders, it is critical to be able to access your folding knife as readily as possible. In this case, the knife should be carried as near to the hands as possible and in such a manner as to allow immediate and unobstructed access to the folding knife.

Referencing accessibility, there are two viable locations for attachment of the modern carry clip. One is the clip attached to the knife and the other is the knife and clip attached to the body of the operator. First let's take a closer look at the clip attached to the knife.

The whole idea of accessible carry locations is proximity to the hands.

Some manufacturers fix the carry clip in only one location and that's all you get – just that one location. If you want to reconfigure the positioning of the clip on the knife handle then you don't have a choice. However, if that's exactly the way you want it and that's exactly the way you intend to carry it, then you don't care!

Other manufacturers design their knives with two positions for placement of the carry clip. In most cases this allows for ambidextrous access. However, there still remain those of us who prefer more options for clip placement on the knife. In this case there are four (4) positions for placement of the carry clip.

Referencing viable locations, there are specific reasons for concern of placement of the carry clip. One of these reasons is accessibility of the dominant hand. If you are a right handed operator, then you'd want to carry your folding knife nearer to your right hand for optimal accessibility. However, if your job responsibilities require you to carry a firearm, that is a holstered handgun to be more specific, and your holster covers your right pocket and you're a right-hand dominant shooter (that is your right hand is your weapon hand – aka "strong side"), then where should your knife go?

Part III

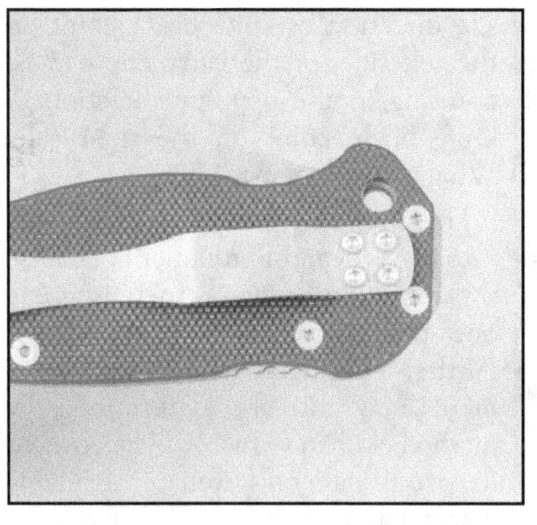
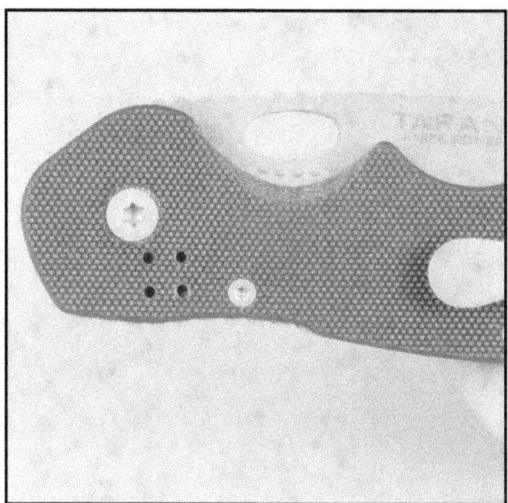
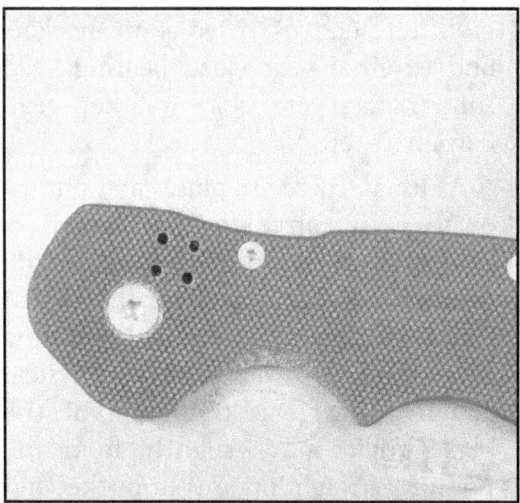
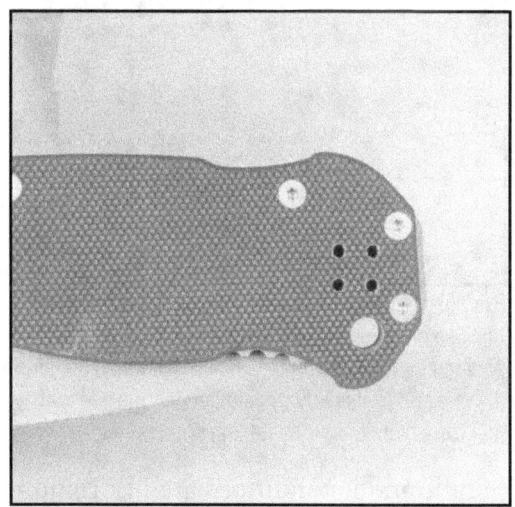

Some manufacturers such as 5.11 Tactical provide four separate positions allowing for maximum versatility of carry clip placement.

This is a very good question. The debate rages on to this day. As we have already studied – there is a plus and a minus for everything and the same holds true for viable carry locations. What about the carry of the knife below the leg holster or even as part of the holster? Again, it comes down to your ability to answer the following questions: What is the intended use for my knife? What is the department policy with reference to carry locations? Will this carry position in any way obstruct my primary responsibilities and access to other equipment based on priority of need?

Another reason for concern of placement of the carry clip is "blade up" or "blade down." In other words the carry clip can be pointing in the same direction of the knife blade point on the left side, or the carry clip can be pointing in the same direction of the knife blade point on the right side or the carry clip can be pointing in the oppo-

Folding Knives: Carry & Deployment

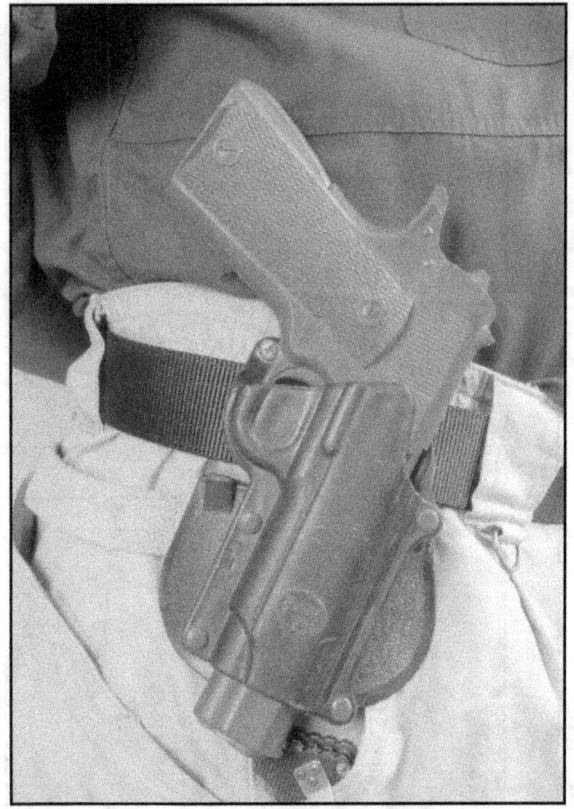

Example of open gun holster on gun belt with folding knife in close proximity of both muzzle and holster. The carry position for the knife on this side of the body interfering with the gun and or its carry system, render this a less-than-optimal carry position for the folding knife.

site direction of knife blade point on the left side, or the carry clip can be pointing in the opposite direction of knife blade point on the right side. What's the difference?

There are many variables that come into play; the most prominent of course is personal preference as there isn't any one that is better than the other. Further consideration of clip placement may also be determined by method of open (which will be covered in detail later) as well as available placement of the blade for carry on clothing such as cargo pant pockets and external gear (load bearing vest, suit coat, severe cold weather gear, carry bags, etc.).

As usual, there are pluses and minuses to each carry location. Based on reports from various students over the years, it has been reported that when carrying a folding knife while running or engaged in other rigorous physical activity, that it is possible that the blade could move slightly from the handle and if conditions are just right and it's pointing straight up in your pocket and you slam your hand in that same pocket to grab the knife quickly, well, although an extreme exception and certainly not the rule, you can do the math.

Other folding knife owners utilizing the carry clip in the exact same configuration as described above have reported nothing but optimal performance. These folks range from first responders to OCONUS war-fighters and use their knives under extreme duress and adverse conditions.

The position of the clip (in any one of the four possible locations) will also determine the carry location on your body which will be covered later in this chapter.

First off the clip should be securely anchored to the knife handle. This can be accomplished in many ways. Glue and pressing are probably not the very best as the tension between the clip and the handle will eventually increase beyond the design of the manufacturer. The very best would of course be welding, and that's great for the one-position clip option – and again, if you're good with that one then that's the absolute best. If not, then the next best thing is screws or similar fastening device(s).

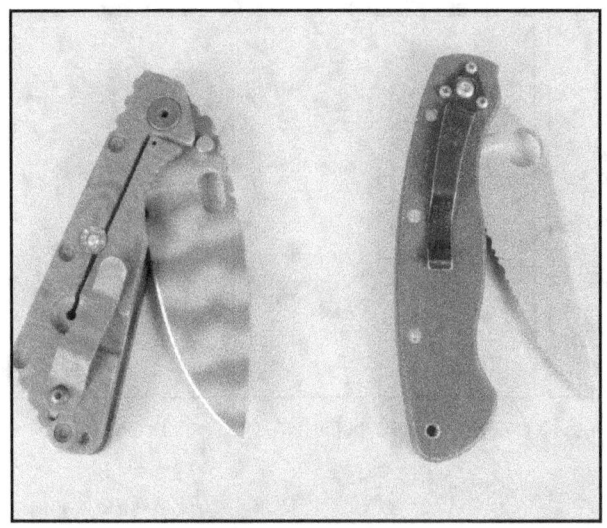

Example of clip down / blade tip up and clip down / blade tip down configurations for right hand dominant application

One or two screws unsupported (that is *without* an additional support tab as part of the carry clip as illustrated below firmly attached to the handle), although sleek and stylish, with repetitive use can eventually weaken and either break the clip off completely or certainly strip the screws and subsequently detach the clip from the handle.

Three screws are usually good enough – although there are some reports of these sometimes loosening over time. A common remedy (by way of preventative maintenance) for loosening clip screws is to apply a drop of "Loc-Tite" or "Vibra-Tite" or some similar self-locking or self-sealing compound (designed for application to threaded fasteners) to the screw thread so as to prevent loosening due to shock or vibration.

The plus to four screws is that it can't get much more secure, but the price tag is it may not look very pretty. The decision again remains with the buyer seeking to balance overall appearance with optimal performance.

The last half of our viable locations is the knife and clip attached to the body of the operator. It is true that the knife can literally be clipped to any part of any gear on the

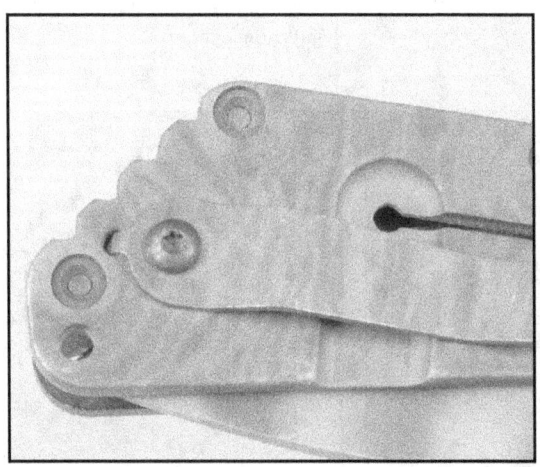

Example of one point secured and supported carry clip. Notice additional reinforcement support tab anchored to handle as described.

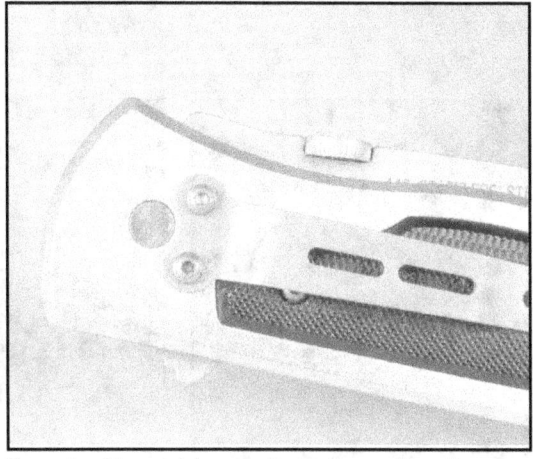

Example of two-point carry clip anchor configuration.

Folding Knives: Carry & Deployment

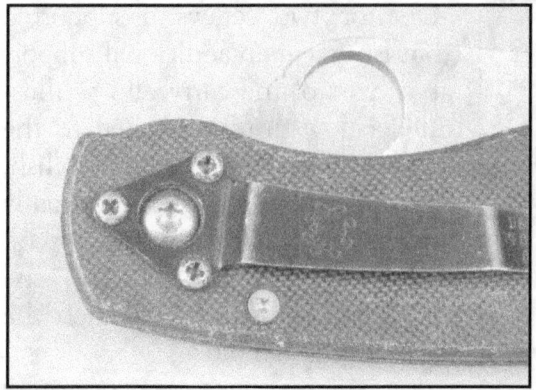

Example of typical three-point carry clip anchor configuration.

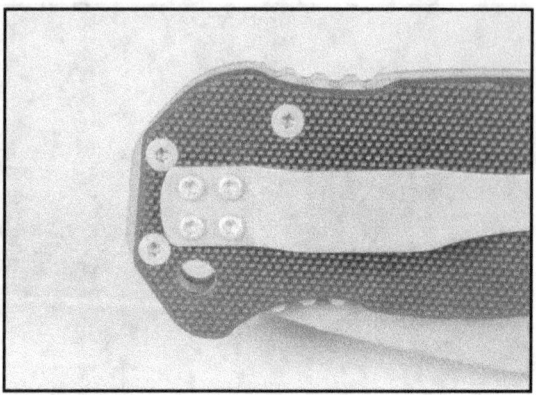

Example of a four-point carry clip anchor configuration.

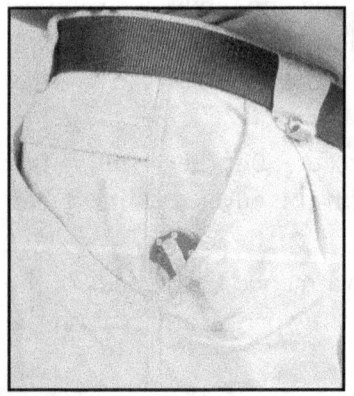

Strong hand (right hand dominant) pocket carry.

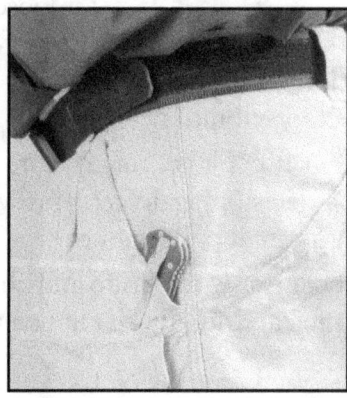

Support hand (right hand dominant) pocket carry.

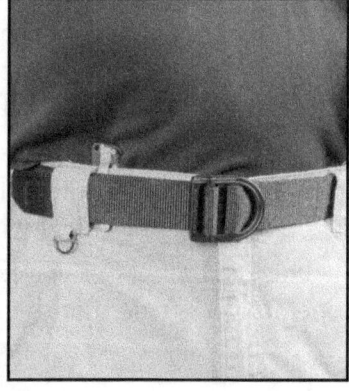

Frontal same-side behind the waistband carry anterior.

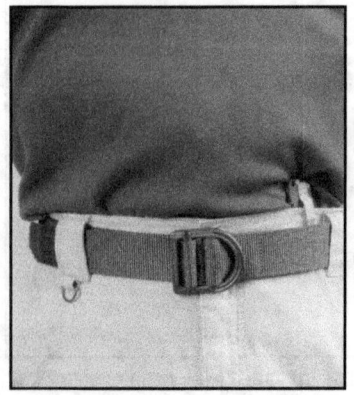

Frontal cross-body behind the waistband carry.

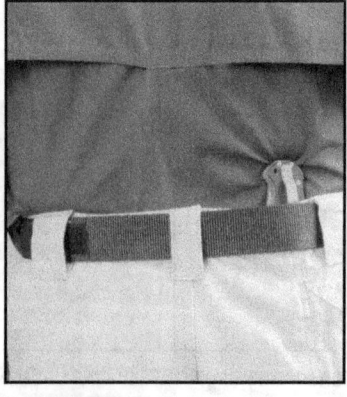

Backside same-side behind the waistband carry.

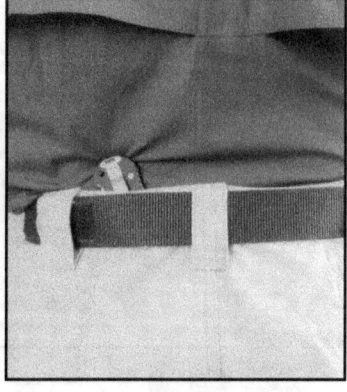

Backside cross-body behind the waistband carry.

body such as load-bearing vests, web-gear, belts, shirt pockets, pants pockets, inside the boot, thigh holster, and even in a sheath slung around the neck. Again, all of these are simply answers to the question: What am I going to use the knife for? However, there are a few more common than others which are universally accepted both per policy and by convention.

The reason why a gun holster is located either on the belt or on the thigh is specifically for accessibility with respect to proximity of the hands. The same reasoning applies to the viable carry locations of a folding knife. The whole idea of accessible carry locations is proximity to the hands.

Locking and Unlocking

Recalling the two essential parts of any folding knife – the blade and the handle, there is an important connection between the two and that is the ability to lock and unlock the blade. The mechanics of locking and unlocking the blade are simple, the blade moves away from the handle (or the handle moves away from the blade) guided by the pivot screw located at the base of the blade and the top part of the handle, from an unlocked and closed position (meaning the knife as it rests "folded") toward an open and locked position.

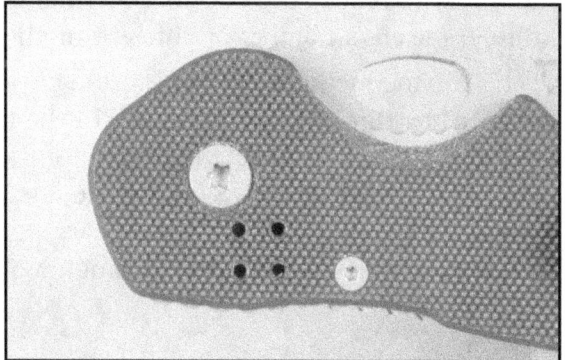
Close-up of a pivot screw which guides the blade.

Potential for personal injury begins as soon as the blade and tip are exposed.

Folding Knives: Carry & Deployment

A side note about pivot screws is that size does matter - the bigger the screw the less chance of its sheering or snapping under opposing handle and blade pressure. Although not a requirement, an oversized pivot screw (as illustrated above) is desirable for any duty folding knife that may inadvertently be used for prying or other rigorous application placing undue stress or tension between blade and the handle.

Paramount safety issues regarding operation of a folding knife are of course safeguarding against unintentional contact with the edge and tip of the blade. With the blade in the folded (or unlocked) position there is little or no risk of injury. However, the split second the blade begins to move from its closed position on the way out to open and, of course, finally in the locked position, is the time when risk of personal injury is greatest.

Varying Methods

In order to reduce the odds of potential injury when opening and closing a folding knife, there are a number of different methods of locking and unlocking the blade.

The varying methods of opening and closing (or locking and unlocking) blades are governed by three component parts of the folding knife which we have already covered in our study: 1. the configuration of the blade itself (shape, size, etc.), 2. type of opening mechanism (wave, post, hole, etc.) and 3. locking mechanism (liner lock, Axis Lock ®, etc.). The terminal learning objective here is to find a method that works between a safe opening (locking) and rapid deployment – the balance between quick and safe.

Safe Open

What the heck is so difficult about opening a folding knife? Everybody knows you just flip it open and you're good to go, right? Well, let's take a closer look at this.

One of the most common methods is to completely bypass usage of the opening mechanism altogether. In this method it is commonly accepted to reach for the folding knife, pull it out from the carry position and with a quick flip of the wrist (which both looks and sounds cool) snap the blade into the open and locked position. This commonly practiced quick wrist flip technique is usually executed with the blade pointing up or the blade pointing down.

Let's first take a look at the blade pointing up "flick" open option.

The remaining "quick flick of the wrist" open option is resulting with the blade in the downward position. Let's take a closer look at this option.

In these first two examples, the plusses are that it takes only one hand to move the blade out from the handle toward the open position and that it initially appears to be very quick (and of course looks and sounds cool).

The minuses of the "quick flick of the wrist" open options are that: 1. the folding knife was not specifically designed (and certainly not under warranty) for this partic-

Part III

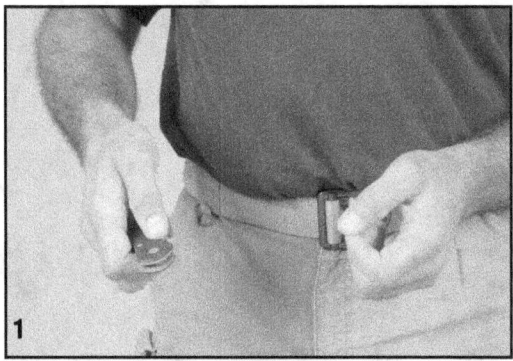

Step 1. Operator firmly grasps knife handle in preparation to open with blade pointing out or up.

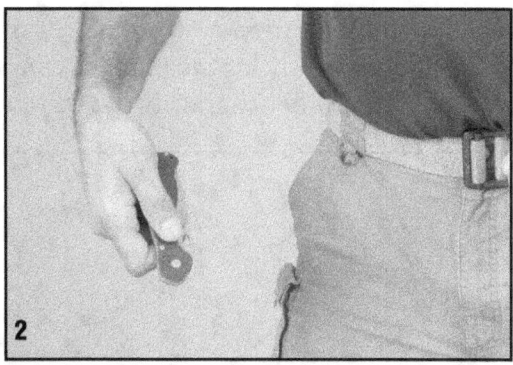

Step 2. Bent wrist initiates opening "flick." Rapidly snapping the wrist causes centrifugal force to take control of the blade.

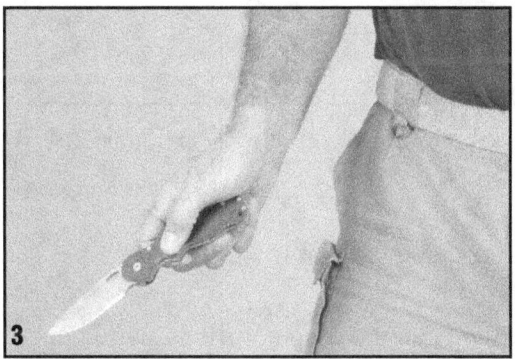

Step 3. Centrifugal force continues to carry the blade to the open and (hopefully) locked position.

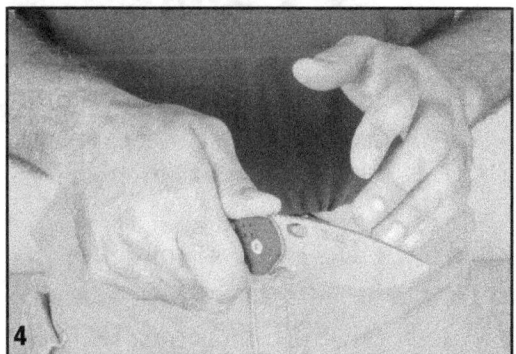

Step 4. After the blade is locked out the operator repositions fingers for a more stable grip on the knife handle.

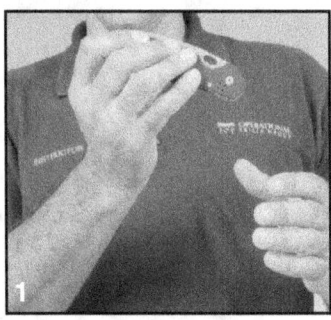

Operator firmly grasps knife handle in preparation to open with blade pointing down or back.

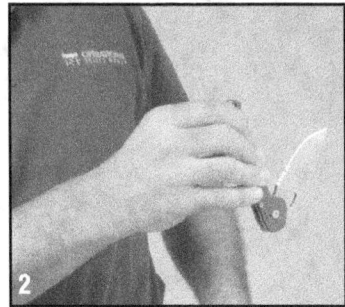

Bent wrist initiates opening "flick." Rapidly snapping the wrist causes centrifugal force to take control of the blade.

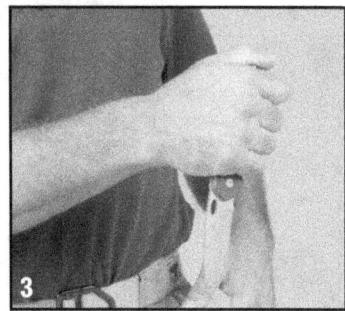

Force continues to carry the blade to the open and (hopefully) locked position. After the blade is locked out the operator repositions fingers for a more stable grip on the knife handle.

107

ular opening method as the knife was manufactured with actual opening mechanism(s) specifically for the purpose of using it to open the knife, 2. there is no control of the blade (other than centrifugal force) and 3. the knife can, under duress, easily

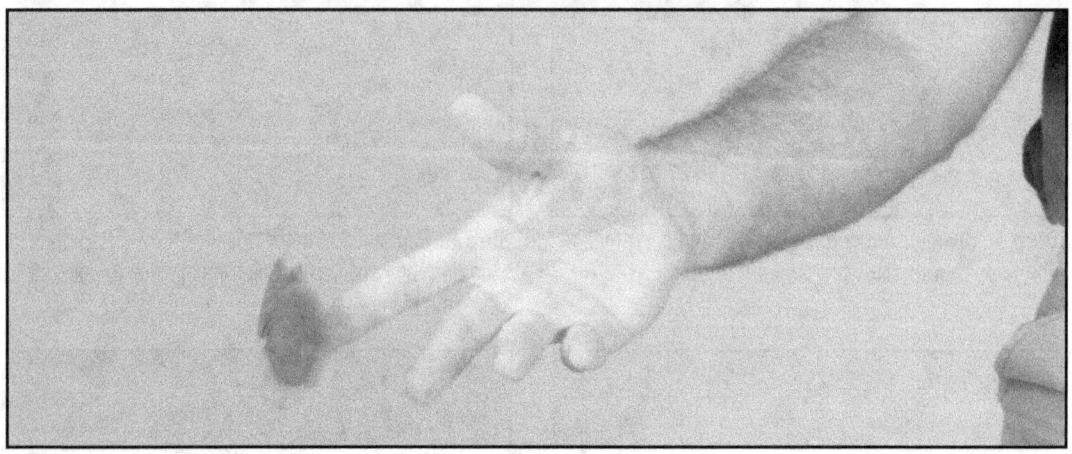

Failure to maintain control of the blade in either of these two common "quick flick of the wrist" open options can result in launching your folding knife.

(and often does) go flying out of the hand completely out of the grip of the operator representing a grave safety risk.

Another method of opening the folding knife is actual usage of the designed opening mechanism and in doing so taking advantage of the features (predominantly safe-

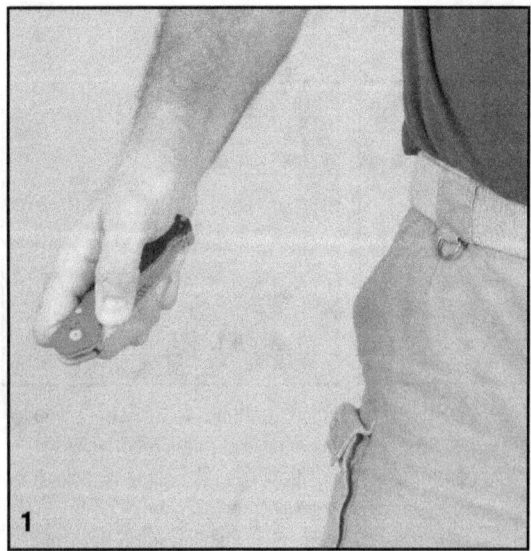

1

Operator places thumb firmly against opening mechanism (in this example the Thumb Hole).

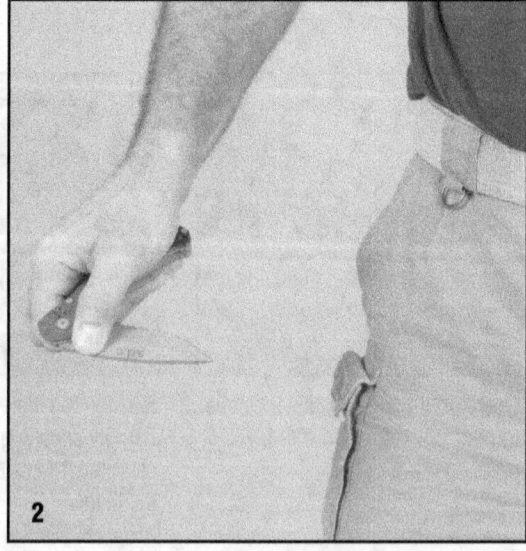

2

Pushing down and away with the opposing digit (thumb) activating the opening mechanism against the fingers (holding the handle) the blade begins to swing safely and in a controlled manner away from the handle.

Part III

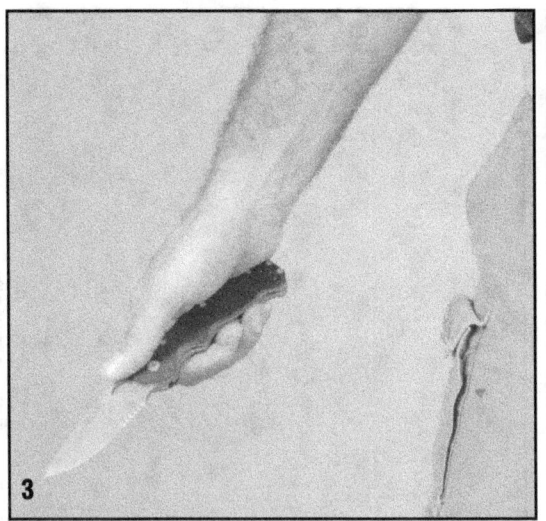
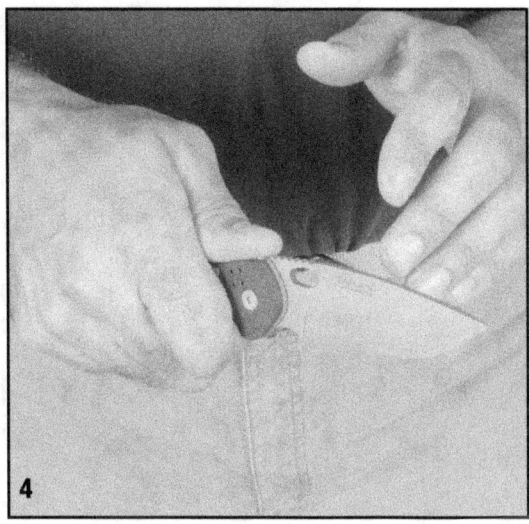

At the end of its opening cycle the blade is moved to the locking position.

After the blade is locked out the operator repositions for a more stable grip on the knife handle.

ty) of that specific opening mechanism. The following is an example of utilizing the Thumb Hole opening mechanism.

Similar to the "flick" opening method, the plusses are that it takes only one hand to open the blade. However, in utilizing the specifically designed opening mechanism of the folding knife blade, there is the added physical and complete control of the blade all along its path all the way to the locked position.

The minus is that it may take about another quarter of a second. Of course there is the option to utilize the opening mechanism for half the opening cycle and then switch over to the flicking method. At least with this method, the blade is in complete control for at least half of the opening procedure.

What I consider to be one of the safest and most secure blade opening procedures is a two-handed method that I developed specifically for professional first responders back in the early 1990's.

Utilizing both hands bring them together as if naturally clapping (this can be executed with your eyes closed or while not even looking at the knife). Next firmly grasp the opening mechanism with your support hand and pull the blade from the handle in such a manner as to end up with your support hand ending up behind your strong hand and with the blade in a stable and locked position facing the edge and tip away from your body.

Tarani Safe Open method:

Utilizing the Safe Open, the blade is under your complete control the entire path.

The plusses include: 1. complete control, 2. simultaneous position of the strong hand into a stable grip (even prior to the blade locking in the open position), 3. maximum support as two hands are stronger than one, 4. rapid safety check capability

Folding Knives: Carry & Deployment

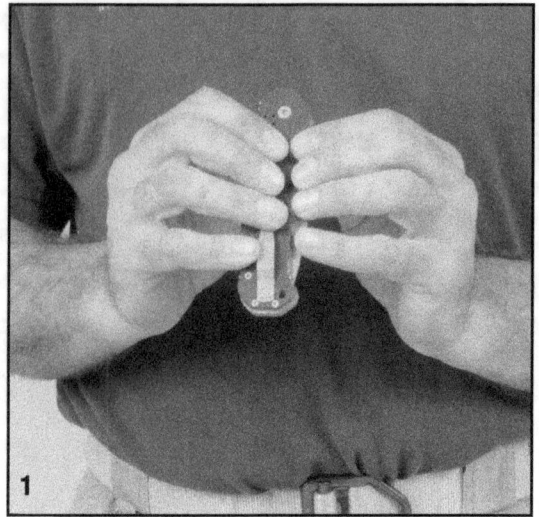

1. Natural hand clapping position (can be executed safely in no or low-light).

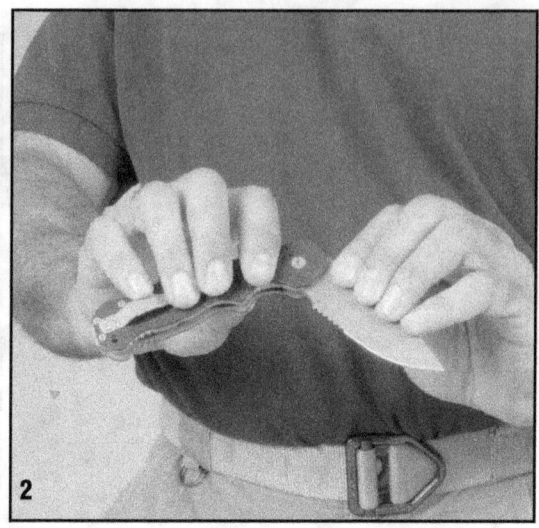

2. Turning the tip and edge away from your body and moving the blade toward the open position in a fully controlled manner

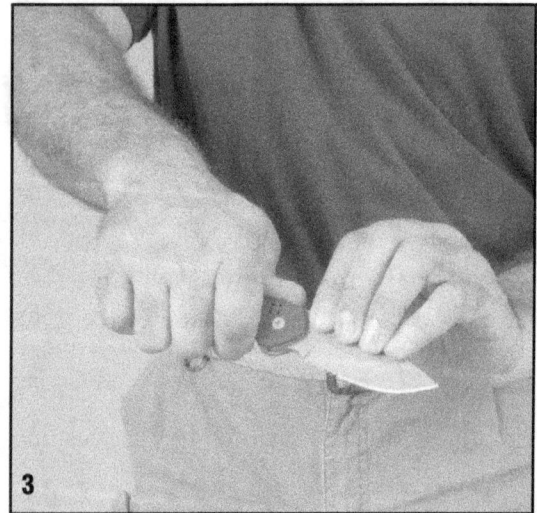

3. Manually test to ensure a stable lock while simultaneously moving strong hand to a more stable and appropriate grip.

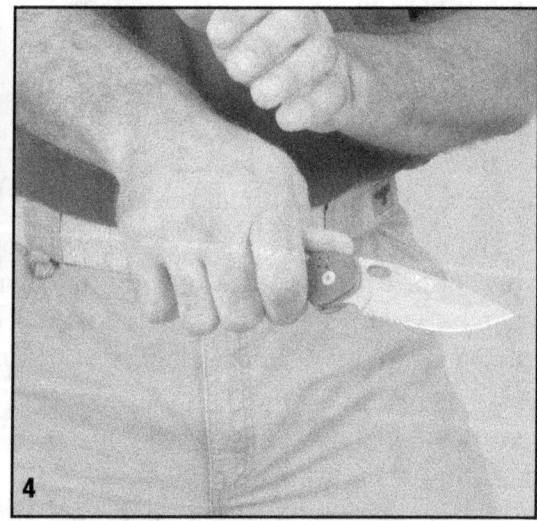

4. Remove the support hand.

(both visual and tactile – especially if low or no-light conditions) to ensure the blade is in fact locked into place at the end of the opening cycle. The minus is that it takes two hands.

In delivery of numerous training classes over nearly two decades (as of this writing), rather than take my word for it, I would always allow my students to "see if it really works" against the clock. In this particular training drill I would ask that they utilize the opening method of their choice such as "the flick," one handed opening mechanism, Safe Open etc., basically any method of their choosing.

Going around the room and on the command of "go" each participant was to open the knife as quickly as possible – but the catch was they had to end up in a stable and locked open position with a safe and stable grip and of course it was timed.

Even against the clock most opening methods (really more the user than the method) were all within hundredths of a second of each other. In many cases the Safe Open method proved to be quicker. Those of you who may have attended some of these classes may recall that speed was never really a consideration – knives flying out of the hand, cut fingers and unstable grips were more the issue!

It is important to mention that no one way to open a knife is "better" than another. In fact there are entire training systems dedicated to the opening and closing of knives – but that's enough information to fill an entirely different volume. Again, the bottom line here is that there is more than one way to open a knife but whatever way is made yours should take into account the aspect of the balance between rapid deployment (which will be covered in greater detail further on) a good stable grip and personal safety.

Back in the days when I was teaching officer knife safety full time, we would line up the entire class in such a manner that there was plenty of space in front, behind and next to each of the attendees for safety. Then, utilizing a stop watch, the class was permitted to pull out their folding knives and open them as quickly as possible – while being timed – about a dozen times or so in a row. The reason we did this was because it's an interesting demonstration when at regular intervals, half-open and fully opened

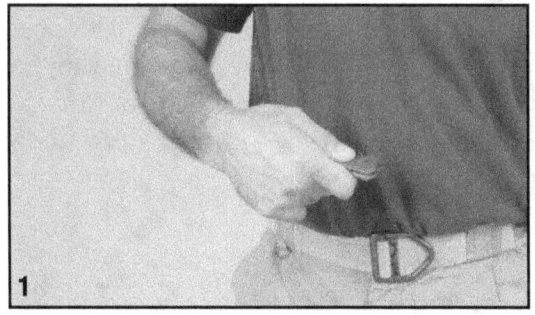
Preparing to flick the wrist to open the blade.

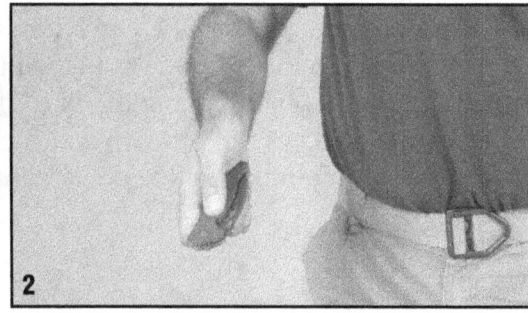
Mid-motion grip remains firm around the handle.

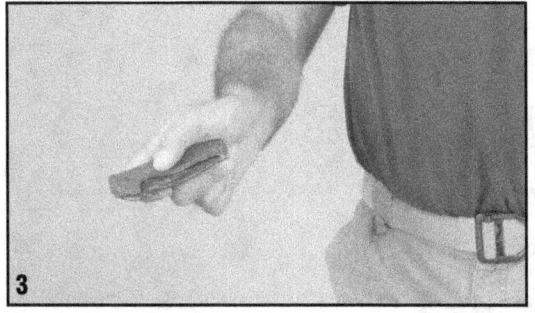
Centrifugal force begins to take control away from grip.

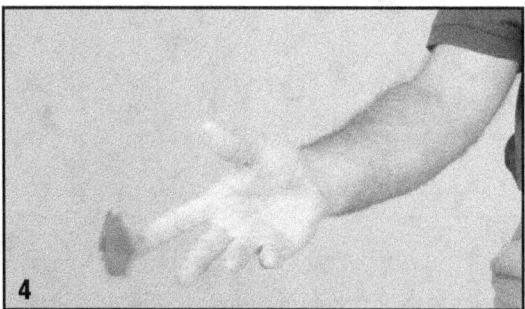
Resulting in removal of folding knife from hand of operator.

folding knives would go zipping out of the hands of their operators and flying across the room. During this type of training an opening sequence going for speed more often than not looked something like this:

Not to say that one opening technique is better than another since everyone's body type varies and of course the endless configurations of blades, opening and locking mechanisms are also a factor, but the desired end result is to effectively and safely open and lock the blade into place without losing control of your folding knife.

The moral of the story here is that even in a controlled environment – a class room – with minimal stress (how intimidating can a stopwatch be right?), and that everyone participating was ready and knew what to expect, things can, and did, go wrong.

Now imagine being out in the field – the freezing cold, where you can barely feel your fingers, or maybe you're wearing gloves, or it's raining, snowing, underwater or other inclement weather or adverse conditions such as extreme duress. Imagine there being other factors involved, life or death pressure, based on the environment or condition of the scenario. Imagine your hand and knife covered in oil, blood or some other viscous material or some combination of all of the above – all of these factors can (and do) significantly impact the "simple" ability to open and close a folding knife.

Locking

Keeping in mind personal safety first (your safety and the safety of those around you) it is advisable that you do not move the blade edge or tip in the direction of a body part – yours or someone else's. Similar to Rule #2 in firearms safety "Never let the muzzle cover anything you're not willing to completely destroy," the same applies to edged tools - when moving the blade from the closed (unlocked) position to the open and locked position, the tip and edge should be controlled in such a manner as to not cover anything that you don't want to cut or poke.

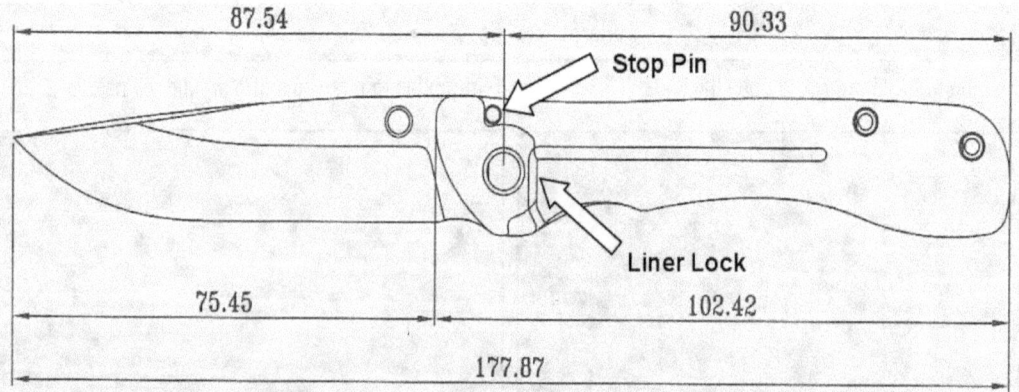

Locking: Notice that both the internal Stop Pin and the Liner lock are pressed flush against the base of the blade and are utilized in such a manner as to secure the folding knife blade into a fixed and "locked" position.

Part III

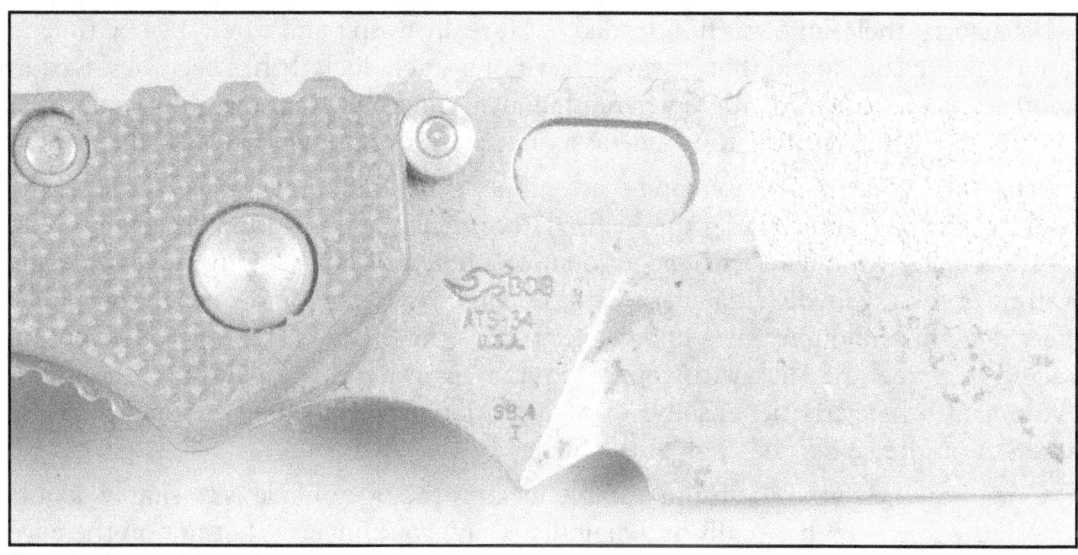

Example of external Stop Pin.

What actually is "locking?" There are (generally speaking) at least two points of contact which are made with the base of the blade which in fact lock the knife blade into place. One of them is called a "stop pin" which can be located either external to the handle (literally attached to the blade) or internal to the handle (as illustrated). The other is the actual locking mechanism itself which can be any one of the plethora of mechanisms available on the market to day – a liner lock is used here for purposes of illustration.

Ensuring a Stable Lock. To open the knife blade from its closed position to the open and locked position requires a sufficient amount of force to engage the locking mechanism as well as pressing the base of the blade firmly against the Stop Pin. It is possible to sometimes miss engaging the locking mechanism.

Referring back to the days of full folding knife training classes of attendees opening and closing knives in a controlled environment one could observe several individuals

Example of blade not fully locked.

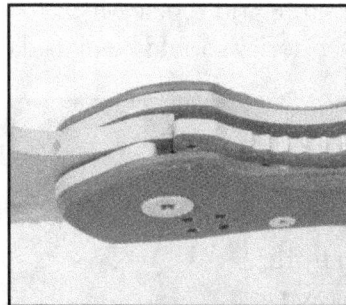

Physically ensure a stable lock (visually and manually).

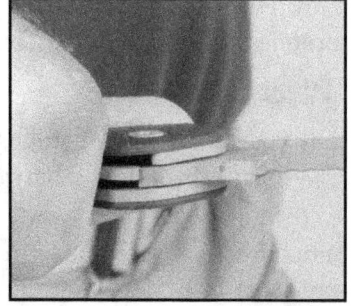

Blade in stable open and locked position.

who thought their knife was full opened but it really wasn't and when it came time to use the knife they found that it was in fact not a stable lock. It has been observed in more than one case in the field where individuals have cut their own fingers on their own knives while opening the blade as a direct result of not ensuring a stable lock.

Most folks go by the audio alone – listening for the ubiquitous "click" sound of the locking mechanism engaging the blade. However, under stress and especially in an operational environment with a considerable amount of noise (sirens, several people yelling or talking loudly, inclement weather, large vehicles moving, etc.), it would be very difficult and often times impossible to hear the "click." The only other option would be to feel the click with your hand. But again wearing a pair of gloves and/or with many other activities ensuing – your attention may be focused on something else and not on the feel of the click of the blade.

One of the best ways to ensure a stable lock on the open blade is to simply double check the lock – both visually and digitally (using your fingers – but not on the edge or the point) quickly test for any blade movement. If it doesn't look like it's locked in place or the blade wiggles around when you test the back of it, then apply just a bit more force and it will lock into place ensuring a stable lock.

Safe Close

Unlike opening the blade – where there are many options (flick of the wrist, one-handed open, two-handed open, etc.) safely closing the blade is very limited in viable options. As a professional instructor I would be remiss in my responsibilities failing to mention the "rapid close." Certain individuals (trained or untrained) tend to want to close the knife faster than they opened it. Take my word for it after observing literally thousands of students, the most unsafe and most effective way to slice open your own digits to the bone is to try and close your knife faster than you open it.

Akin to holstering a gun faster than you can draw it out of the holster, there are certain inherent problems. One of them is speed is placed as a higher priority than personal safety. By placing emphasis on the "looking cool" factor and in trying to generate as much speed as possible, one runs the increased risk of sustaining personal injury.

When it's time to close the knife and replace it in its original carry position there's no need for speed, operational priority should be placed on a safe and controlled close.

Unlocking

If you recall what it takes to lock the blade into place (internal or external stop pin and locking mechanism – in that order) the reverse of this is what it takes to unlock. The very first step is to disengage the locking mechanism. By

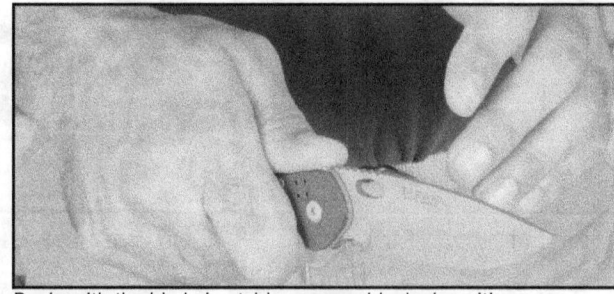
Begin with the blade in stable open and locked position.

Part III

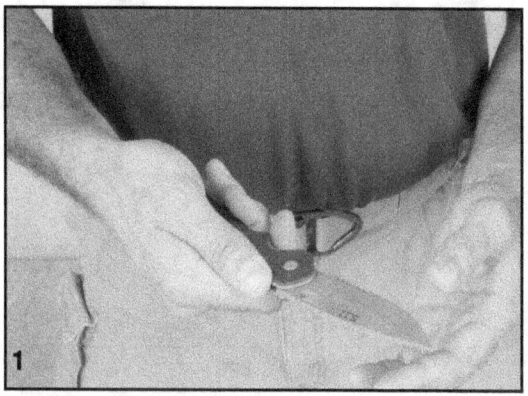

Turn the folding knife in such a controlled manner as to easily access locking mechanism (a liner lock is shown in this example).

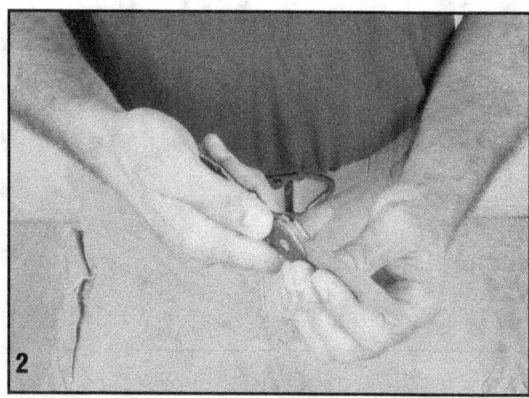

Take full control of open and locked blade with both hands – one remaining on the handle and the other on the back side of the blade.

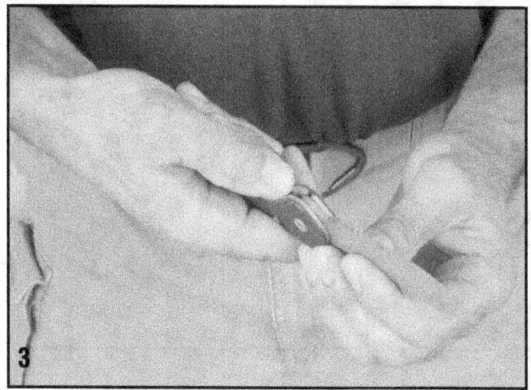

Disengage the locking mechanism (a liner lock is shown in this example).

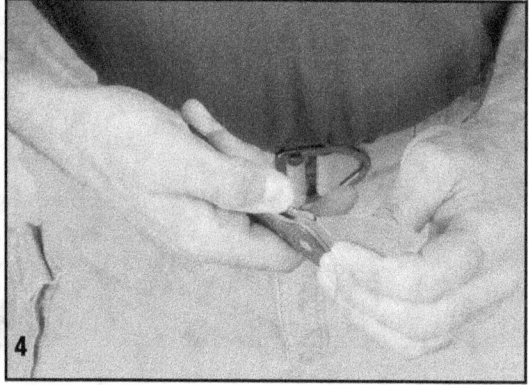

Using the hand controlling the blade apply sufficient force to "break" the lock.

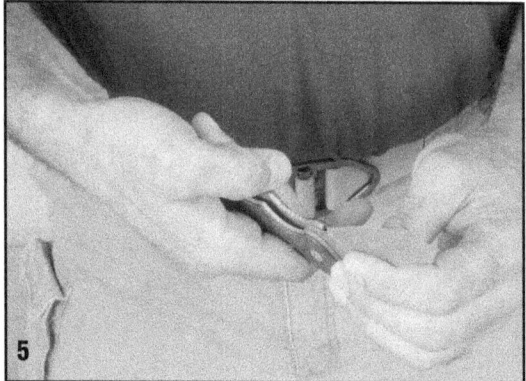

Remove any and all appendages away from the path of the inbound blade.

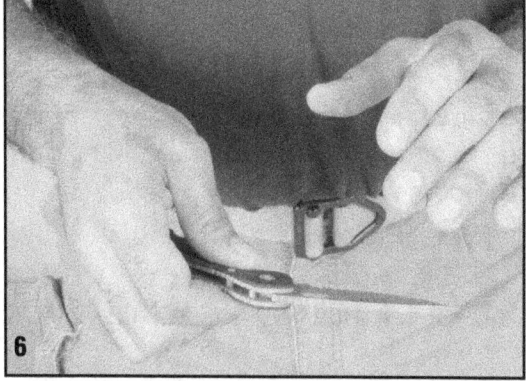

Begin to turn the partially open blade edge and tip away from your body (and anyone in your immediate area of operation). The highest potential for injury is when the blade is not fully locked or secured inside the handle as the blade edge and tip remain free-floating.

Folding Knives: Carry & Deployment

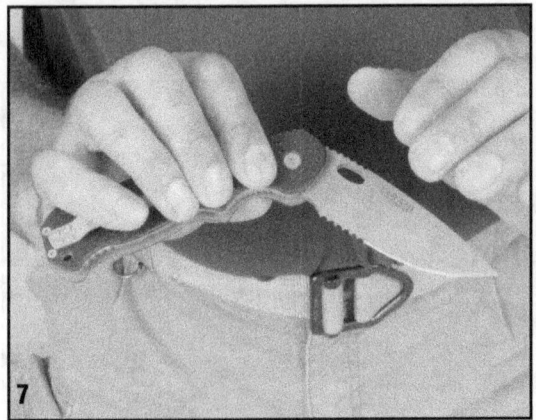

7. Place your free hand directly behind the partially closed blade with the palm facing in the same direction as the blade edge.

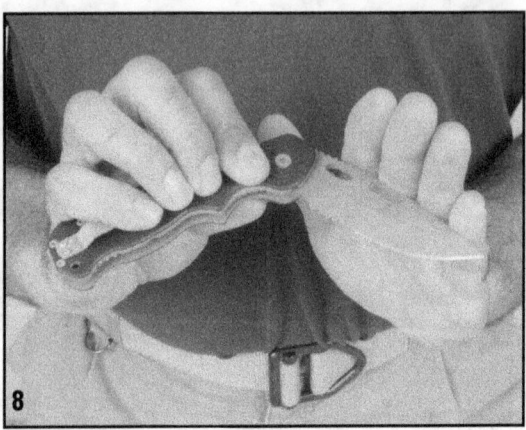

8. Maintaining a clear path between the blade edge and the handle, apply force to the back of the blade.

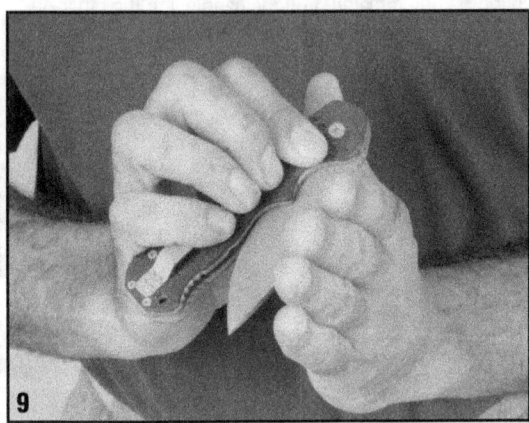

9. Continue moving the blade toward the handle.

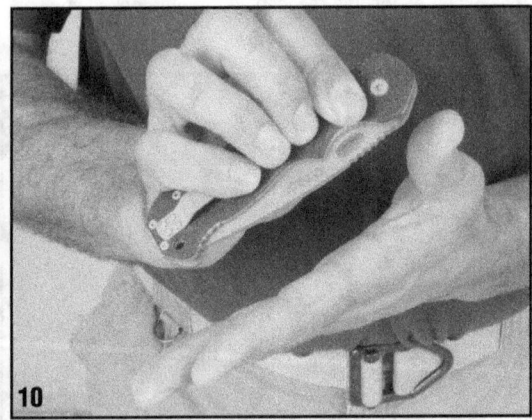

10. Until a Safe Close is achieved.

disengaging the locking mechanism, this allows the blade to move freely in the opposite direction of the stop pin.

Ensuring a Safe Close

As pedantic as it may sound the safest method of closing any folding knife is to ensure that nothing (especially your fingers or any other appendages) is placed along the path of the blade as it is moving from the open and locked position to the closed (unlocked) position.

In order to ensure a safe close the following steps are strongly recommended:

Getting a Grip

Similar to the usage of any hand-held tool, the folding knife is no exception to the importance of establishing and maintaining a solid grip on the handle.

The human hand is designed with four fingers and an opposing digit. The opposing digit works best when applying force forward and in opposite direction of the force applied by the fingers. In technical terms these would be "intersecting force vectors," but for the rest of us silver-backs it's what keeps a hammer or a can of beer from falling out of our hands.

The same concept applies to gripping a folding knife. It is critical that the force of the opposing digit (thumb) be applied opposite the force of the fingers. The design of the handle and friction radii of most quality production folding knives meet this important requirement.

Determining Your Grip

Purchasing a solid grip on the folding knife is critical to positive control of the blade and overall handling of the folding knife. Some knife owners have bigger hands while other knife owners have smaller hands, others have long thin fingers while still others have five sausages sticking out the end of their arm.

We are all unique in our body types and sizes and therefore there's no such thing as a folding knife (or any other hand-held tool for that matter) that's a "one size fits all," especially with regards to grip. This is why if you choose to be proficient with your knife that you will be able to determine what fits best for you. In other words, in the knife world there's no such thing as determining "the" grip, but quite the opposite – determining your grip.

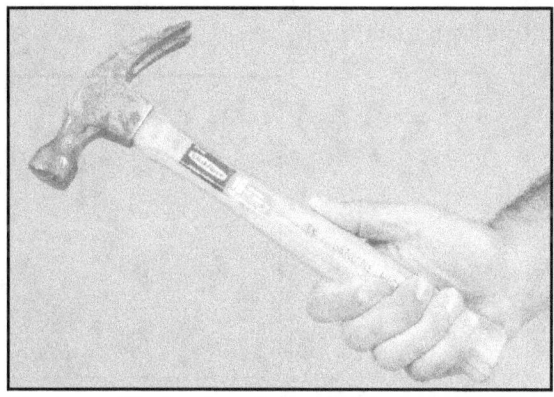

The force of the thumb applied against the force of the fingers is what holds the tool in place.

Specific Grips

The knife is a fairly versatile tool and can be held virtually anywhere you can place your hand. Holding the blade is NOT recommended. The one and only rule of thumb (no pun intended) is maintaining this opposing digits (intersecting force vectors) concept as per above.

Overall there are two types of grip predominantly applicable to the folding knife. One grip allows the blade tip to be positioned forward and the other the blade tip in reverse position.

The last two grips illustrating tip positioned in the downward position are predominantly for use with the blade in digging or scraping motion.

As mentioned above (see section on Folding Knife Selection), ergo dynamics plays an important part in the election of specific grips. Knife designers (much like car designers) continually seek to incorporate the balance of ergo dynamics with functionally in that they want a final product that both fits the body comfortably yet can deliver when it comes time for the end product to function. Even though the folding knife may be designed with ergo dynamics in mind it may not be designed to fit your

Folding Knives: Carry & Deployment

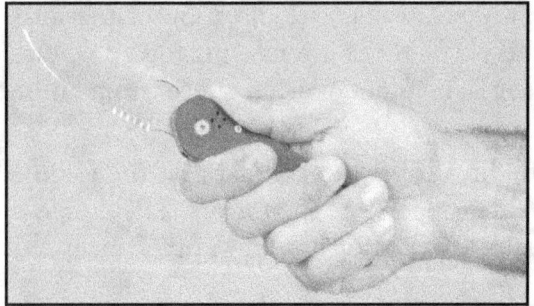
Grip with blade tip forward – conventional usage.

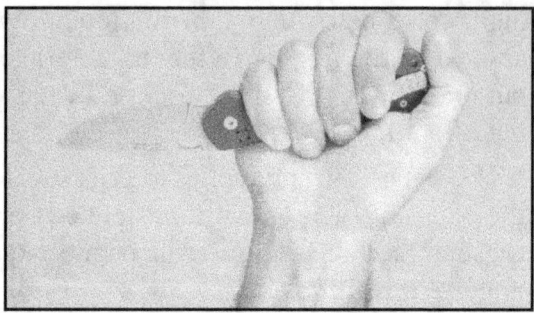
Grip with blade positioned downward edge facing out.

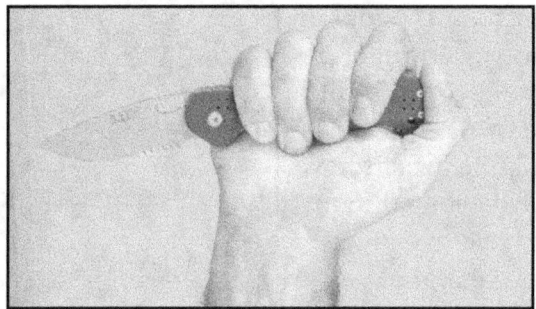
Grip with blade positioned downward edge facing in.

specific grip. This is one of the main reasons for the plethora of folding knives on the market today.

Given that most knives these days function as utility tools predominantly, most modern production folding knives are designed to be utilized with the blade in full locked and open position (obviously for usage of the blade) as well as closed or unlocked position perhaps with peripheral attachments such as a bottle opener, window breaker tool or Strike Plate.

Optimal Grip Open. The modern production folding knife is carefully designed with not only ergo dynamics in mind but also with functionality as the primary intended goal.

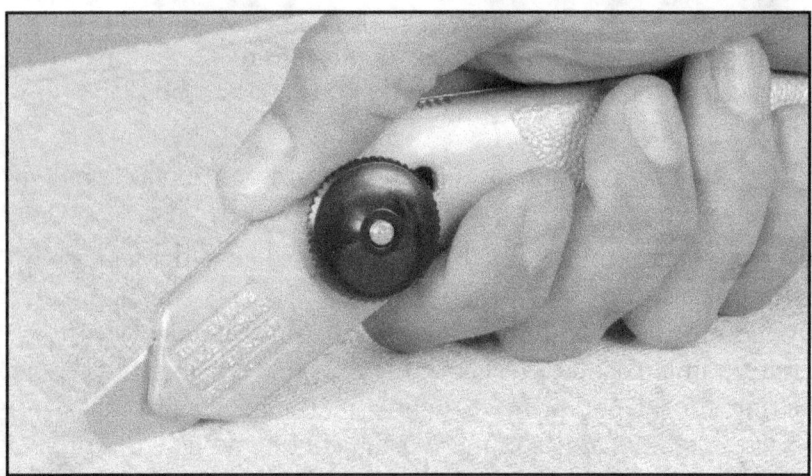
Typical drywall cutting knife. Designed specifically for usage of the cutting edge – not the tip. Notice the shape of the handle and the placement of the fingers and palm of the hand in the grip – specific to continuous cutting motion.

digit works best when applying force forward and in opposite direction of the force applied by the fingers. In technical terms these would be "intersecting force vectors," but for the rest of us silver-backs it's what keeps a hammer or a can of beer from falling out of our hands.

The same concept applies to gripping a folding knife. It is critical that the force of the opposing digit (thumb) be applied opposite the force of the fingers. The design of the handle and friction radii of most quality production folding knives meet this important requirement.

Determining Your Grip

Purchasing a solid grip on the folding knife is critical to positive control of the blade and overall handling of the folding knife. Some knife owners have bigger hands while other knife owners have smaller hands, others have long thin fingers while still others have five sausages sticking out the end of their arm.

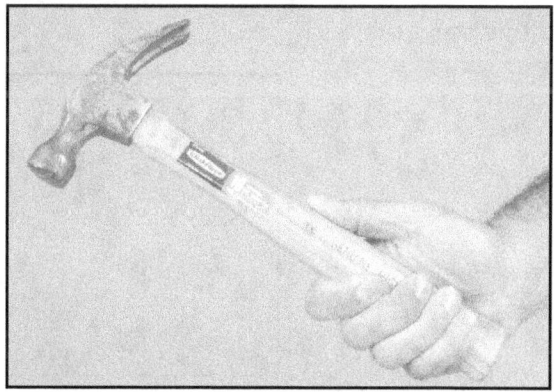

The force of the thumb applied against the force of the fingers is what holds the tool in place.

We are all unique in our body types and sizes and therefore there's no such thing as a folding knife (or any other hand-held tool for that matter) that's a "one size fits all," especially with regards to grip. This is why if you choose to be proficient with your knife that you will be able to determine what fits best for you. In other words, in the knife world there's no such thing as determining "the" grip, but quite the opposite – determining your grip.

Specific Grips

The knife is a fairly versatile tool and can be held virtually anywhere you can place your hand. Holding the blade is NOT recommended. The one and only rule of thumb (no pun intended) is maintaining this opposing digits (intersecting force vectors) concept as per above.

Overall there are two types of grip predominantly applicable to the folding knife. One grip allows the blade tip to be positioned forward and the other the blade tip in reverse position.

The last two grips illustrating tip positioned in the downward position are predominantly for use with the blade in digging or scraping motion.

As mentioned above (see section on Folding Knife Selection), ergo dynamics plays an important part in the election of specific grips. Knife designers (much like car designers) continually seek to incorporate the balance of ergo dynamics with functionally in that they want a final product that both fits the body comfortably yet can deliver when it comes time for the end product to function. Even though the folding knife may be designed with ergo dynamics in mind it may not be designed to fit your

Folding Knives: Carry & Deployment

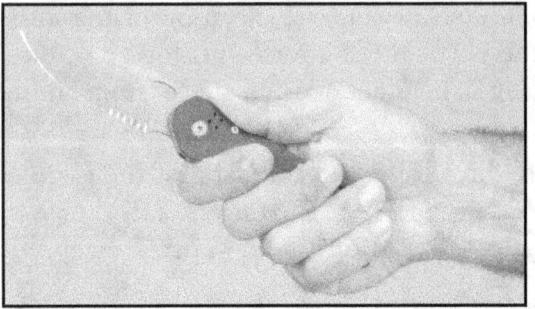
Grip with blade tip forward – conventional usage.

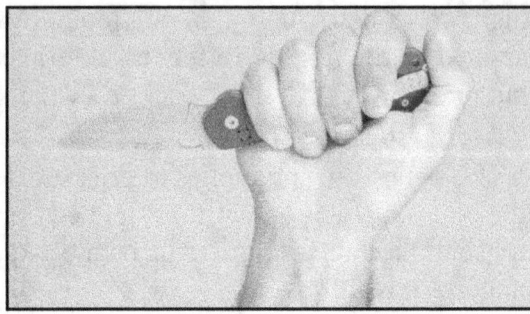
Grip with blade positioned downward edge facing out.

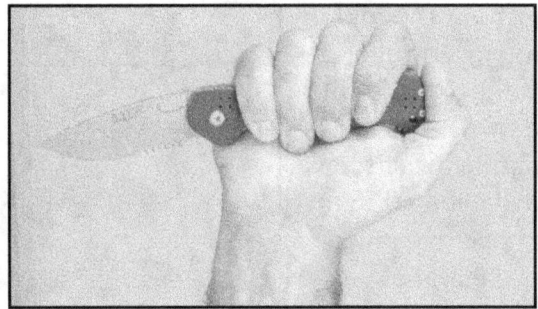
Grip with blade positioned downward edge facing in.

specific grip. This is one of the main reasons for the plethora of folding knives on the market today.

Given that most knives these days function as utility tools predominantly, most modern production folding knives are designed to be utilized with the blade in full locked and open position (obviously for usage of the blade) as well as closed or unlocked position perhaps with peripheral attachments such as a bottle opener, window breaker tool or Strike Plate.

Optimal Grip Open. The modern production folding knife is carefully designed with not only ergo dynamics in mind but also with functionality as the primary intended goal.

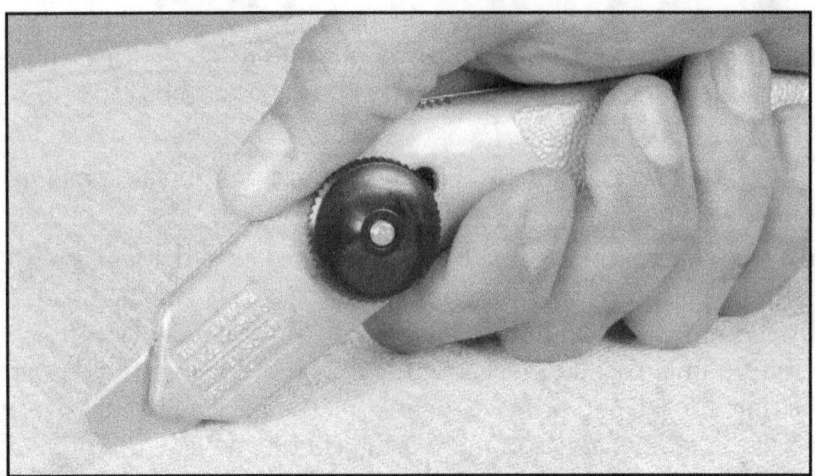
Typical drywall cutting knife. Designed specifically for usage of the cutting edge – not the tip. Notice the shape of the handle and the placement of the fingers and palm of the hand in the grip – specific to continuous cutting motion.

Such functionality can include both cutting (utilizing the cutting edge) and puncturing (utilizing the tip). Most folding knife designers allocate equal "grip ratio" applied to both the cutting function (edge) and the thrusting or puncturing function (tip).

If you look at a common construction cutting tool such as a drywall knife or a box top cutter, the primary function of that tool is cutting (usage of the cutting edge) only. Any attempts to use utilize the very tip of the razor blade for prying or puncture would result in a broken tip.

In order for it to be fully functional, the folding knife must allow the user to grip the knife in the blade open position in such a manner as to remain functional for both cutting (edge) and thrusting (tip). To facilitate this functionality there are generally a set of friction points, friction ramps, finger stops or "knurling" (as described above – see Friction Points) which can be found on the knife handle and in some cases on the blade.

To facilitate usage of the cutting edge the friction radii are strategically placed at the top of the handle (near the base of the blade), machined as part of the trigger finger

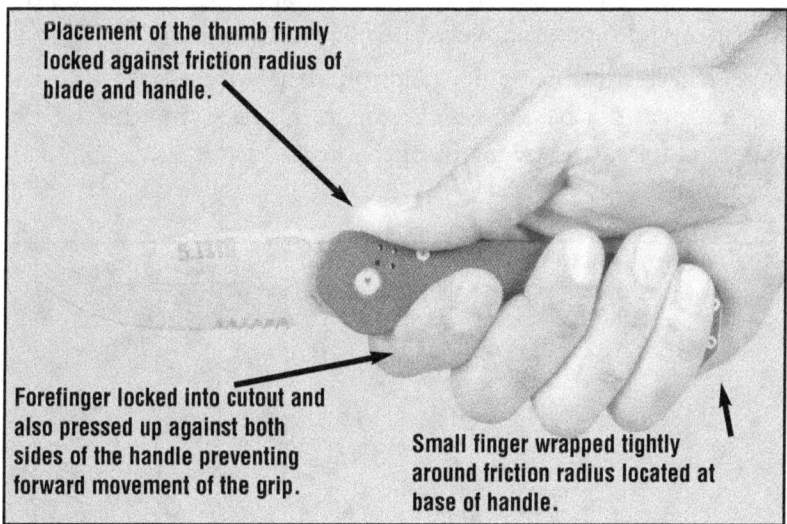

Placement of the thumb firmly locked against friction radius of blade and handle.

Forefinger locked into cutout and also pressed up against both sides of the handle preventing forward movement of the grip.

Small finger wrapped tightly around friction radius located at base of handle.

When executing a cutting or chopping movement, placement of the thumb, forefinger and pinky finger firmly against the friction points prevents the movement of the blade from a positive grip.

cutout (where the liner lock is located) and at the bottom part rear of the handle just in front of or at the lanyard hole. The combination of these three points is what secures the folding knife from peeling out of your grip when utilizing the sharpened edge for cutting and chopping movements.

The same is true for the exact opposite movement of the blade. Opposite movement of cutting, chopping or slicing is thrusting or prying which is utilizing the tip in a forward and forceful motion.

To facilitate usage of the tip or point of the folding blade, friction radii are strategically placed in positions at the top of the handle (near the base of the blade and in some cases as part of the actual back of the blade), machined as part of the trigger finger cutout (where the liner lock is located) and at the top part rear of the handle which adds more of a "bite" into the palm of the hand when utilizing the tip.

Folding Knives: Carry & Deployment

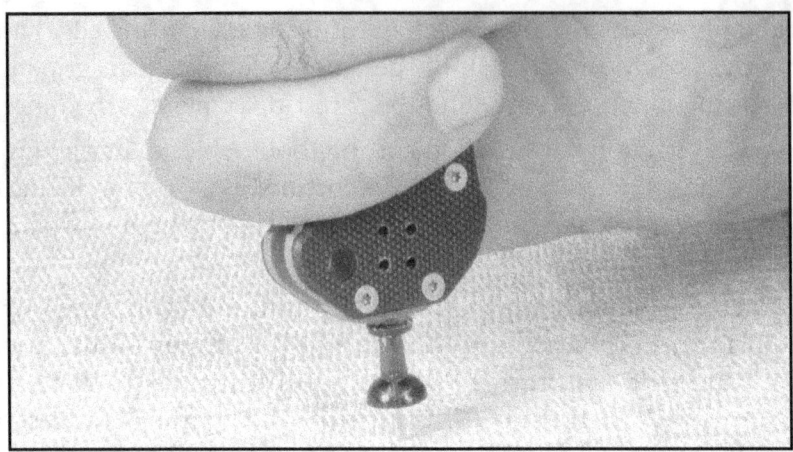

An example of using a Strike Plate – notice positive grip of folding knife in the closed position with the assist of friction points.

The combination of these three points is what secures the folding knife from peeling out of your grip and splaying your fingers and hand wide open when utilizing the tip or point for thrusting or prying type of movements.

Referencing the usage of these friction points (friction radii) when operating the folding knife their importance cannot be expressed enough in preserving the flesh in a sudden thrust or unexpected forward movement of the knife blade.

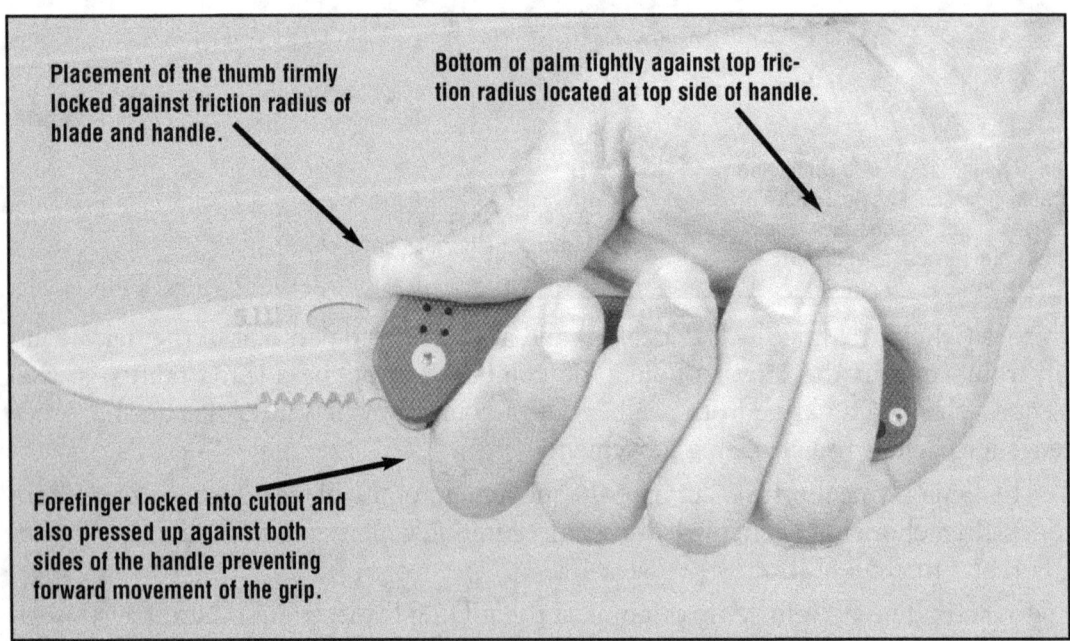

Placement of the thumb firmly locked against friction radius of blade and handle.

Bottom of palm tightly against top friction radius located at top side of handle.

Forefinger locked into cutout and also pressed up against both sides of the handle preventing forward movement of the grip.

When executing a thrusting or prying aggressive forward movement, placement of the thumb, forefinger and meaty part of the palm firmly against the friction points prevents the movement of the blade from a positive grip up onto the sharpened part of the steel.

Part 3

Optimal Grip Closed. As was mentioned earlier, there are times when the folding knife may be utilized with the knife blade located in the closed and unlocked position (bot-

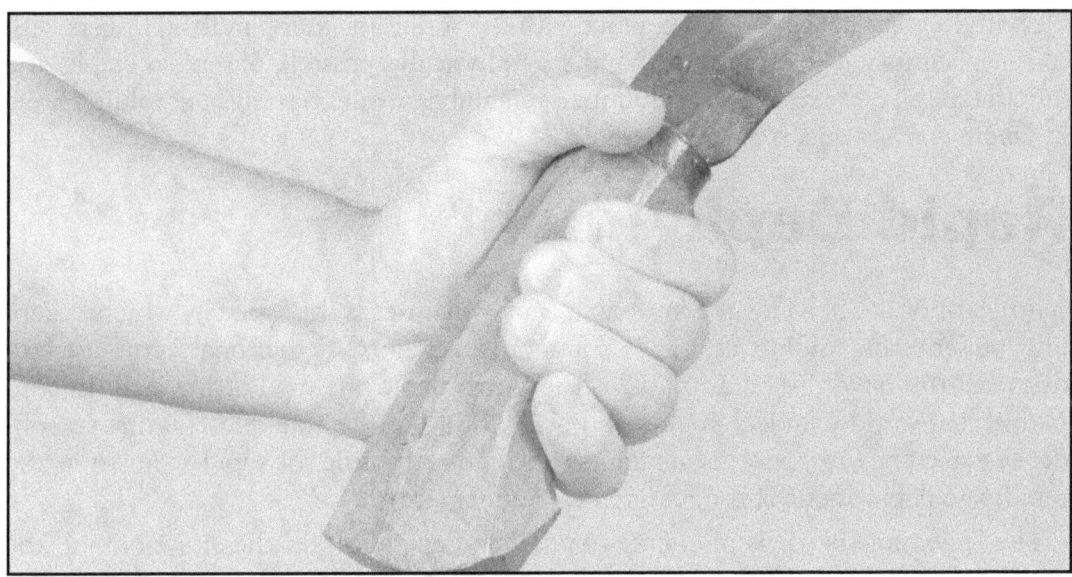
A grip with less surface contact reduces control of the knife blade.

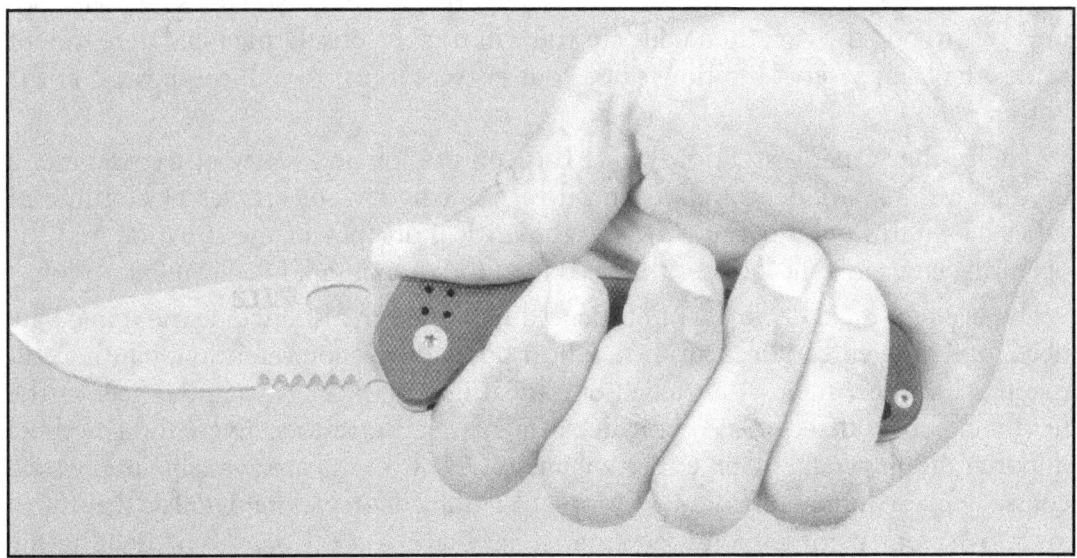
A grip with more surface contact increases control of the knife blade.

tle opener, window breaker, Strike Plate, etc.). Usage of the folding knife in the folded configuration demands that he blade remain in the closed and unlock position throughout. An optimal method of securing the blade into the folded position while acquiring a positive grip on the handle is to wrap the fingers safely around the closed dull backside of the blade further securing the blade into the closed position.

Grip is arguably the most important part of safely handling any folding knife blade whether in the open and locked position or in the closed and unlocked position.

The secret to a sound, positive grip is to make and keep as much contact as possible with the handle. Maximum contact with the surface of your hand and fingers with as much contact with the knife handle as physically possible ensures a stable and functional grip. A simple rule of thumb to remember about gripping your folding knife is: "more contact equals more control."

Rapid Deployment

Inherent of the need for the knife in a timely manner is the speed by which to rapidly present the folding knife. A commonly accepted operational term used to describe "the rapid presentation of" is the word "deployment." Putting it all in context of real-world practical application, we will utilize the example of a first responder or emergency-response team member on scene requiring the rapid acquisition and rapid, safe deployment of the folding knife under stress.

The "need for speed" is of course completely dependent on the situation and the environment in which you may be operating. If it's your day off from work on a lazy summer day and you're out fishing with some buddies sitting back in the shade under a tree half asleep and reaching for another imported beer and you're using your folding knife to pop the cap, then odds are you will not be considering rapid deployment (unless its been a very long time since your last vacation) to pull those caps off the bottle.

Change the scenario and now you're back on the job and you're at the scene of a multiple-vehicle accident and you need to immediately cut articles of clothing or other materials from a burning victim, the critical urgency of the situation and the operating environment dictates the need for a functional rapid deployment option.

The need for rapid deployment of the folding knife can also include the immediate need for self-defense, officer survival, officer safety or weapon retention, but unfortunately all of these topics – although of critical importance to the line officer in the field – are considered taboo with department administrators. Even though these important topics could fill an entire volume on its own (and are available in physical course delivery format, click on www.opskillsgroup.com for more details), this information is very purposely omitted from this text in order to prevent debilitating aneurisms up-chain. As such we will limit our scope of study only to the utility-based operational aspects of the folding knife.

The component parts of any rapid deployment of a folding knife include the combination of all that we have covered up to now including identification, familiarization and working knowledge of the parts of a folding knife, the selection of an appropriate folding knife, optimal carry location(s) and, of course, a safe open and safe close of the folding knife blade. Rapid deployment is simply a matter of putting all of these together and in a safe and controlled manner with quickness.

Presenting the Folding Knife

Similar to firearms, specifically the handgun, the technical terminology slung around the knife community is "presentation," that is the process of locating the folding knife in its carry position, acquiring a stable grip, removing from carry position and safely opening the knife in such a manner as to not adversely impact yourself or your environment. In many cases a first responder, as per our model example above, has very little time to cut away at seatbelts, shoelaces, or even cut someone down from a beam (attempted suicide) – these are only a handful of reasons why a folding knife may need to be safely deployed from its carry position and in a timely manner.

In a real world example (courtesy of an officer who shared the details of this incident in an Officer's Knife Safety class delivered in the state of Indiana), it was a hot summer day in the middle of August in the late 1990's when these officers received the call of an attempted suicide. Code three all the way they burst through the front door and quickly scanned the room, looking up they see the victim holding on to his throat with both hands. Moving their eyes upward they notice that his entire body is suspended by a leather belt tied to a beam running across the ceiling.

As more responders poured through the front door the words "who's got a knife? Were repeated a number of times. While two of the officers now sweating profusely, scramble to set up a platform to try and reach the victim, patting down their gear in response to the knife question, not one of those first responders in that room on that particular day had a knife of any kind in their possession – fixed or folding. The seconds ticked away and by the time the paramedics got there it was too late. Had even a single folding knife been present on scene that incident may have ended differently.

Deployment Training Drills

Second only to having a knife clipped to your body in an accessible carry location, is the ability to physically deploy the knife in a rapid and controlled manner.

Back in the late 1990's in response to a request (issued to the Operational Skills Group, LLC) from a certain high-profile US government agency, I was tasked with the tall order of creating and beta-testing a training package that would assist in the development of "physical deployment skills with a folding knife" for field agents.

Why was it such a tall order? The criteria for this request was very specific in that the training needed to be brief (very few steps – as minimal as possible), concise (right to the point – no fluff and no superfluous movement), simple (keep it simple - easy to learn), geared toward agents with literally no background whatsoever in the usage of

Folding Knives: Carry & Deployment

knives (other than to cut a steak or to butter toast), and easy to assimilate in a brief period of time.

Once the program was developed it needed to be beta-tested on actual students which took some time. After delivery of several classes (and with the kinks all worked out) part of the overall Officer Knife Safety training package included a comprehensive block of study on the development of rapid deployment skills. In this next section, matching all the criteria as per above, the following training drills are still used to this very day as part of this same training course.

Drill One

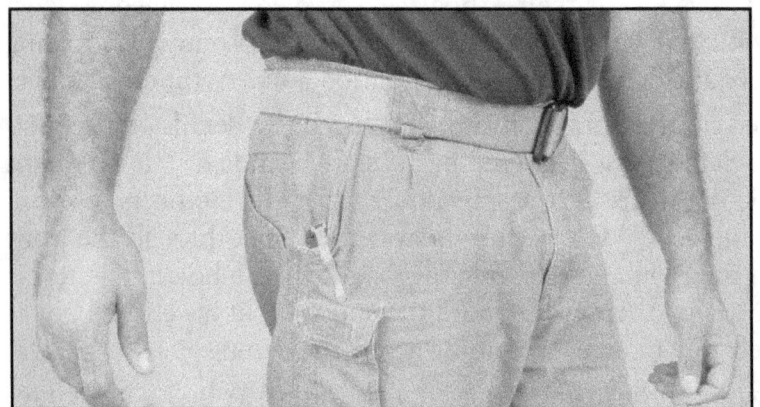

Start with both hands relaxed below your belt.

To set up for this drill stand facing forward with your knife in its normal carry location. Start with both hands relaxed below your belt. Note: for purposes of this training drill the term for your dominant hand is "strong hand" and the term for your non-dominant hand is "support side."

Step 1. Grip – All in one motion move both hands up from below your belt in such a manner as the support hand moves toward the center of your body and position the strong hand in such a manner as to acquire a positive grip on the folding knife as it rests in the carry position.

The movement should be smooth, not jerky or broken and the hands should move simultaneously – not one before the other. Think of flipping a light switch and when you hit that switch both lights click on at the same time. Similarly you want to move the hands both at the same time when you "hit that switch." This first step is called the "Grip" step.

Step 2. Once a positive grip is purchased of the folding knife located in its original carry position, the next step is to clear the folding knife from its carry position. This second step is called the "Clear" step. Included as part of this step is sneaking your thumb over to the opening mechanism ("T"-stop, pin, hole, etc.) and prepare for the next step.

Part III

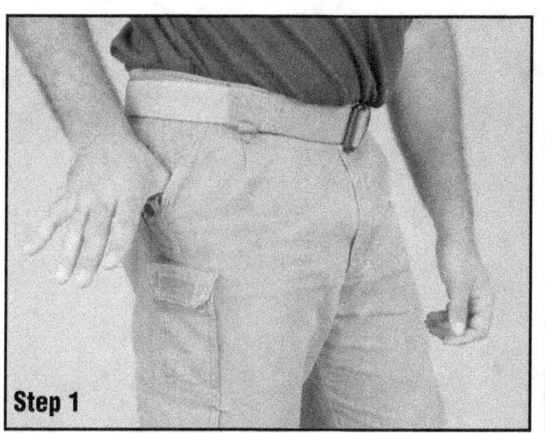

Step 1
Reach thumb first into the pocket.

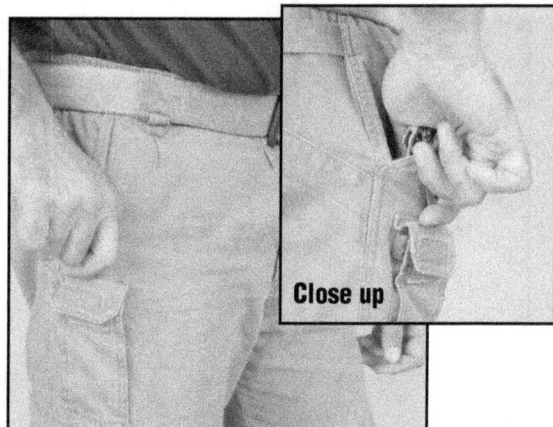

Secure a positive grip on the handle - this is the "Grip" step.

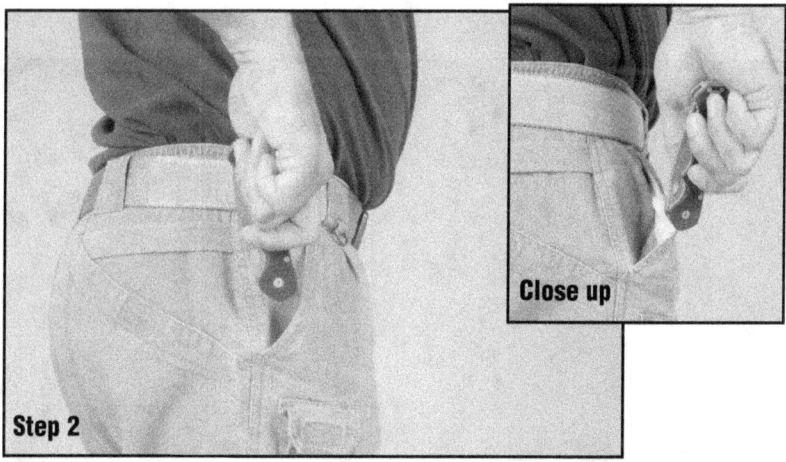

Step 2
Move to the second position – "Clear" the knife from its carry position.

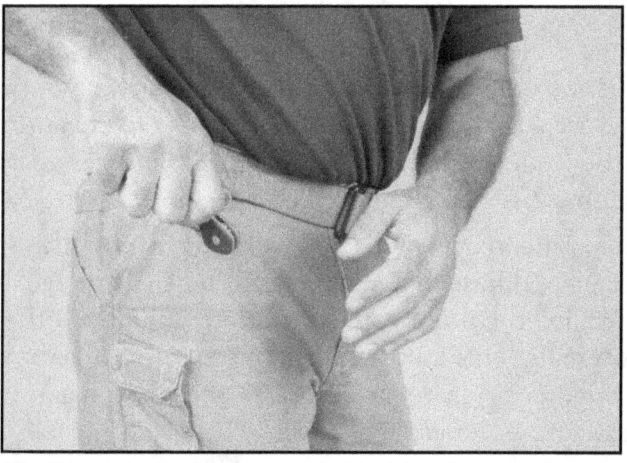

Simultaneously clearing the carry position place the strong hand thumb firmly against the opening mechanism.

Folding Knives: Carry & Deployment

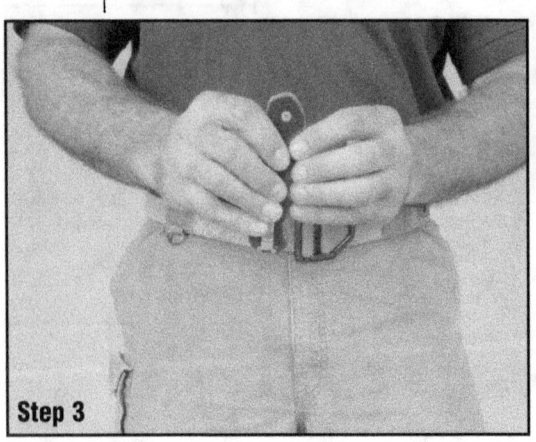

1. Begin the first step of the Safe Open method by bringing both hands together in the natural hand-clapping position.

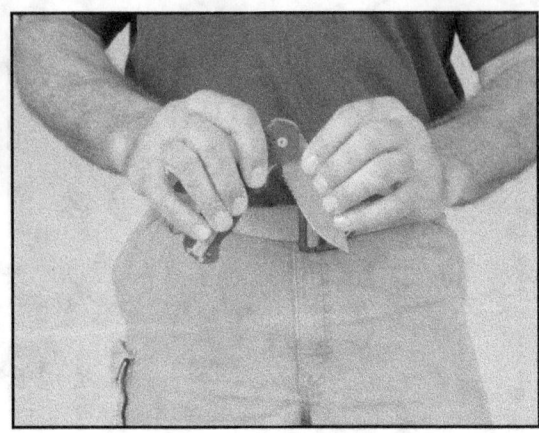

2. Utilizing the Safe Open method, begin to draw the blade from the handle in a safe and controlled manner.

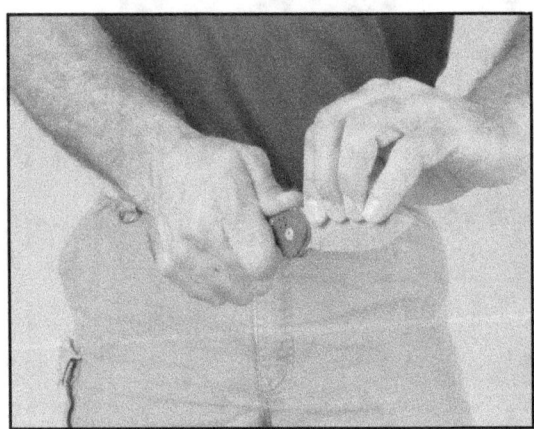

3. Move to the third position – "Safe Open."

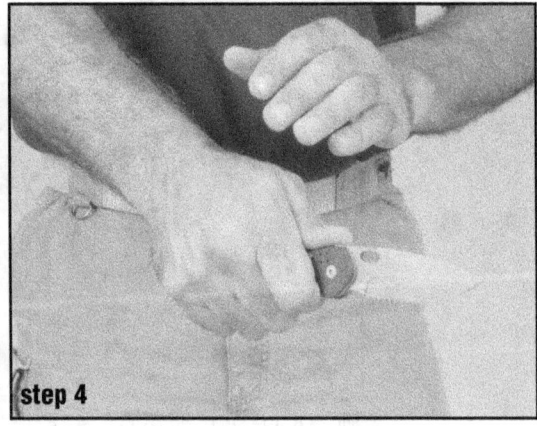

4. Move to the fourth position – "Ready."

Step 3. Once the knife is clear and with the strong side thumb already in position, this allows the operator any number of opening methods (one hand, two hands, etc.) however for consistency we will continue with the Safe Open. Executing a Safe Open (as per above) bring the knife blade from the unlocked and closed position to the open and locked position. This step is called the "Safe Open" step.

Step 4. Once the knife is in an open and stable locked position, this step is to adjust your grip with the strong hand in such a manner as to take advantage of the design of the folding knife and placing the thumb on the friction radius and other fingers along their respective friction radii. This fourth and final step also calls for your feet to be in a stable and supportive position to safely utilize the knife blade in a controlled manner. This step is called the "Ready" position.

To reset the drill simply execute a Safe Close as per above (see Ensuring a Safe Close) - Begin with disengaging the locking mechanism: Once the lock is disengaged, remove any digits from the path of the blade. Next, turn the blade edge and tip away from your body (and anyone in your immediate area of operation). Utilizing the palm of your support hand bring the blade to a safe closed position.

As with all training the number of quality repetitions determines the level of skill. It is strongly recommended to execute this four-step process at least twenty (20) repetitions per training cycle. A training cycle could be while watching TV and you might get up and run the drill during commercials.

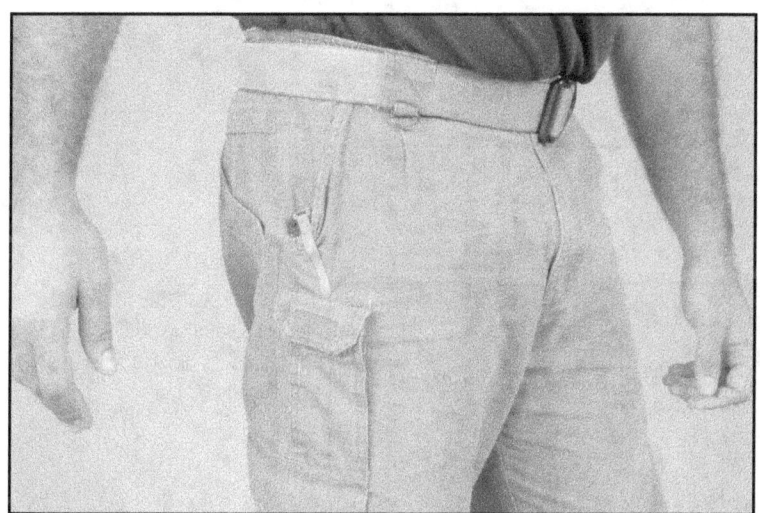

Start with both hands relaxed below your belt.

Drill Two

The next drill in this sequence is to reduce the number of steps by 50% but maintain the same smooth movement and actions minus any robotic movement.

Similar to Drill One, set up for this drill standing facing forward with your knife in its normal carry location. Start with both hands relaxed below your belt the exact same as Drill One.

Step 1. Grip and Clear – Combining Steps 1 and 2 of Drill One above, all in one motion move both hands up from below your belt in such a manner as the support hand moves toward the center of your body and position the strong hand in such a manner as to acquire a positive grip on the folding knife as it rests in the carry position. Once a positive grip is purchased of the folding knife located in its original carry position, the next step is to clear the folding knife from its carry position.

Step 2. Safe Open and Ready - Combining Steps 3 and 4 of Drill One above, once the knife is clear and with the strong side thumb already in position, this step is simply executing a Safe Open by bringing the knife blade from the unlocked and closed position to the open and locked position. Adjust your grip with the strong hand in such a manner as to take advantage of the design of the folding knife and placing the thumb on the friction radius (friction stop) and other fingers along their respective friction stops.

Folding Knives: Carry & Deployment

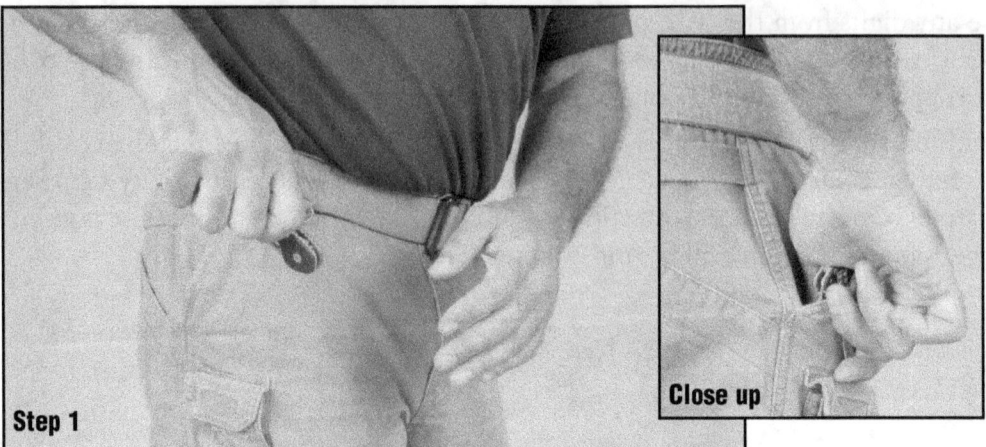

Step 1. Execute the first and second steps of Drill One as one fluid motion — "Grip and Clear"

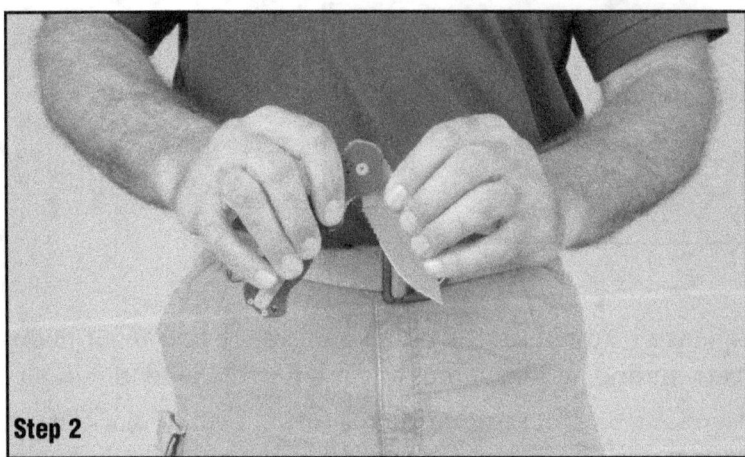

Step 2. Execute the third and fourth steps of Drill One all as one fluid motion — "Safe Open and Ready"

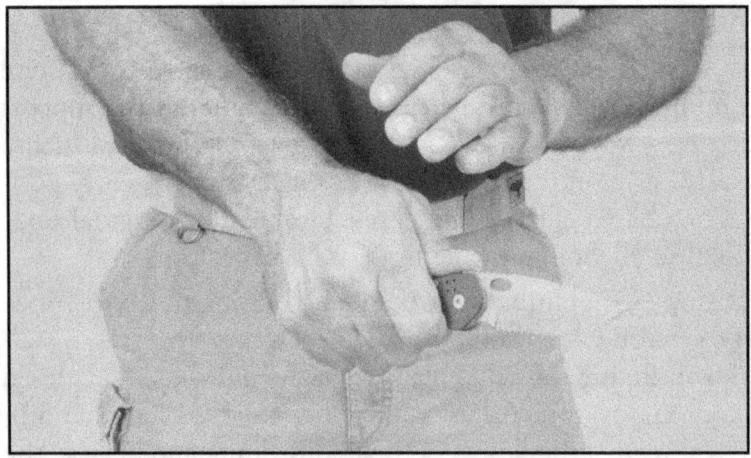

Part I

To reset the drill simply execute a Safe Close as per above (see ensuring a Safe Close) - Begin with disengaging the locking mechanism: Once the lock is disengaged, remove any digits from the path of the blade. Next, turn the blade edge and tip away from your body (and of course anyone in your immediate area of operation). Utilizing the palm of your support hand bring the blade to a safe closed position.

Drill Three. The next drill in this sequence is to reduce the number of steps down to a single fluid motion, but maintain the same smooth movement and actions minus any robotic movement.

Similar to Drill One, set up for this drill standing facing forward with your knife in its normal carry location. Start with both hands relaxed below your belt the exact same as Drill One and Drill Two.

Step 1. Grip, Clear, Safe Open and Ready – Combining Steps 1, 2, 3 and 4 of Drill Two above, all in one motion move both hands up from below your belt in such a manner as the support hand moves toward the center of your body and position the strong hand in such a manner as to acquire a positive grip on the folding knife as it rests in the carry position. Once a positive grip is purchased of the folding knife located in its original carry position, the next step is to clear the folding knife from its carry position.

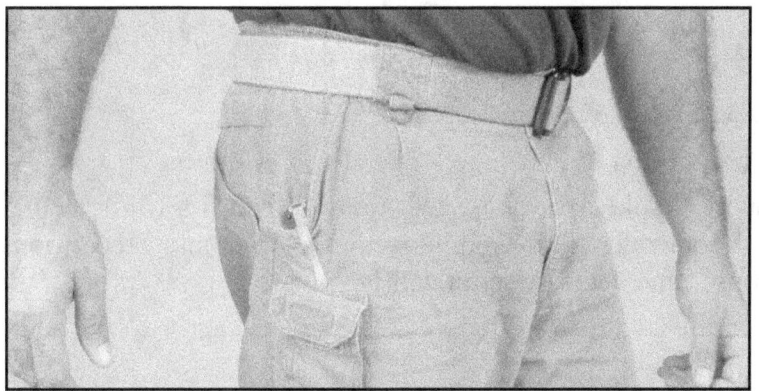

Start with both hands relaxed below your belt.

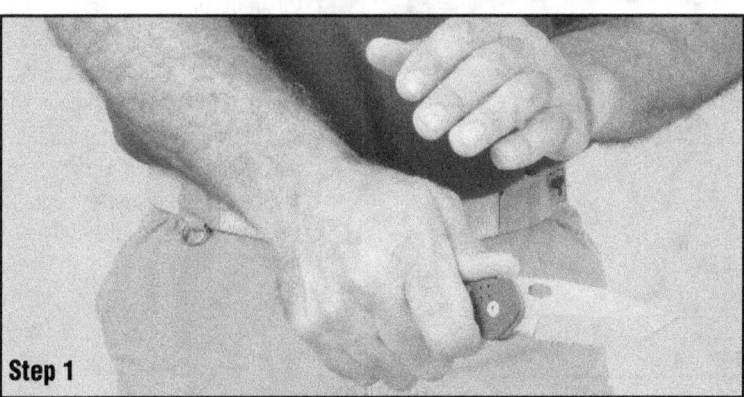

Step 1. Execute Steps 1 and 2 of Drill Two combined bringing all of the deployment steps together – "Grip, Clear, Safe, Open and Ready" as one fluid motion.

Folding Knives: Carry & Deployment

Once the knife is clear and with the strong side thumb already in position, this step is simply executing a Safe Open by bringing the knife blade from the unlocked and closed position to the open and locked position. Adjust your grip with the strong hand in such a manner as to take advantage of the design of the folding knife and placing the thumb on the Friction Radius and other fingers along their respective friction stops.

To reset the drill simply execute a Safe Close as per above (see Ensuring a Safe Close) - Begin with disengaging the locking mechanism: Once the lock is disengaged, remove any digits from the path of the blade. Next, turn the blade edge and tip away from your body (and of course anyone in your immediate area of operation). Utilizing

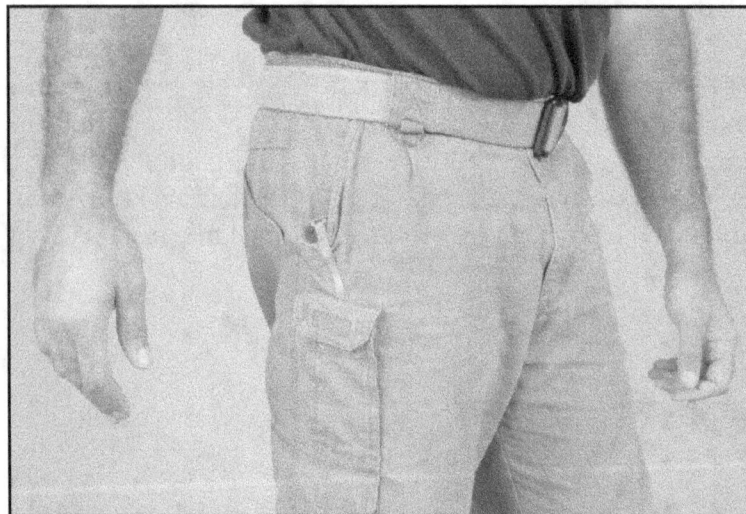

Start with both hands relaxed below your belt. Have the time keeper standing off to the side (not in front of you and your knife hand just in case the knife goes flying out of your hand during the execution of the drill) with stopwatch in hand at the ready to issue commands.

the palm of your support hand bring the blade to a safe closed position.

As time goes on and you would like to decrease your response time (as this is a rapid response drill), you may ask someone to time your movements. Over many repetitions you will notice your response time decrease significantly.

Drill Four. Setup for this drill is identical to the previous three except with the addi-

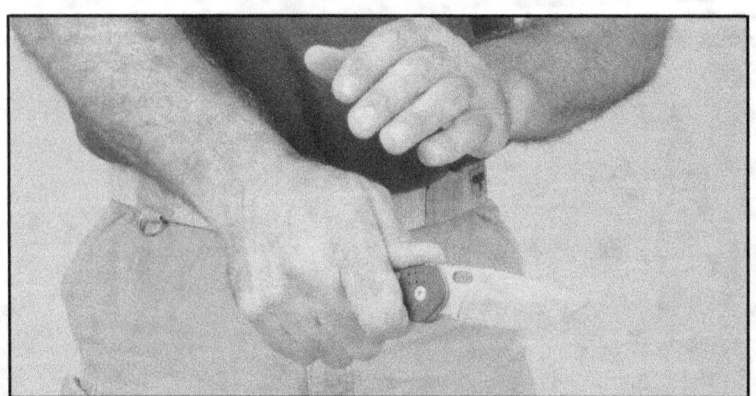

When the time keeper asks if you are ready and issues the command of "Go" this is your cue to execute Drill Three as rapidly and smoothly and in as a controlled manner as possible. Upon completion of your movement the clock will be stopped and the time read aloud.

tion of a timekeeper and a stopwatch.

Step 1. Timekeeper says "Are you ready?" You nod signifying that you are ready.

Step 2. Timekeeper upon observing that you are ready issues the command "Go" and simultaneously clicks the stopwatch activating the timer. Immediately upon hearing the command to start, execute Drill Three in a timely manner. When the timekeeper observes that you are in a stable and locked ready position, he will stop the running time.

Step 3. After hearing your time execute a Safe Close (as per above) record your time on a piece of paper.

Repeat this drill at least ten times in order to start engraining memory and motion.

The Doctor's Wife

Anyone can execute these drills – it just takes a little time and plenty of repetitions. In most training classes, operators were deploying their folding knife in faster times than they would be able to deploy their handgun.

In one of the many Officer Knife Safety classes delivered over the years, I seem to recall it being in the Mid-West. There was a combination of officers and SWAT team members as well as the team physician and his wife (the doctor's wife). It was approved by upper chain that she could participate in the class as (if memory serves) she was somehow part of hospital staff. On the first day of running these drills her times were something like three and sometimes as much as five seconds on the timer. All of the alpha males and of course her husband (the team physician – who literally patted her on the head) smiled and said that all she needed to do was practice. She was irate. The instructions were to practice that night and we'd re-run these same drills in the morning and try for new response times.

The next morning the doctor came into class all bleary-eyed clutching two cups of coffee. His wife looked like the terminator with her fierce expression of determinism. The doc looked like he'd been up all night in the operating room or perhaps working overtime. When asked, the doc said "No, it was my wife – she didn't sleep, all night long I heard her in the bathroom practicing her rapid deployment drills." He gave me a glaring look and said, "Click, click, click, click, click, all night long that's all I heard – couldn't get a minute of sleep in!" Everyone laughed – until the running of the drills.

On the clock and with everyone watching in amazement, she consistently hit sub-second presentations (less than 1 second response time from carry position to full presentation). The class fell silent. Even the quickest guys in the class couldn't touch her score. The doc was not having a happy day.

Folding Knives: Carry & Deployment

Perishable Skills

Although simple to execute and easy to get up on these skills, what goes up must come down. Without continual refresher training or maintenance repetition training these skills will diminish – never completely gone, but certainly measurably diminished. As with all skills (swimming, running, weight lifting, shooting, etc.) if you don't use them – you lose them. It's an unfortunate fact of physical training.

Position and Balance

Now prepared with the background and knowledge of Folding Knife history, Blade Geometry, Metallurgy, parts of the Blade, parts of the Handle, Accessibility, Carry Positions, Locking, Unlocking, Safe Open, Safe Close and Rapid Deployment, the final training in operational proficiency is both in safe handling and usage of the folding knife in conventional and unconventional positions.

The two most important elements of effectively working with any folding knife are where your body and hands are in relationship to what project you are working on (position) and the even distribution of your body weight in such a manner as to effectively and safely operate the knife blade (balance). In other words, the key components to safe and controlled usage of the knife blade are position and balance. Let's take a closer look at each component individually.

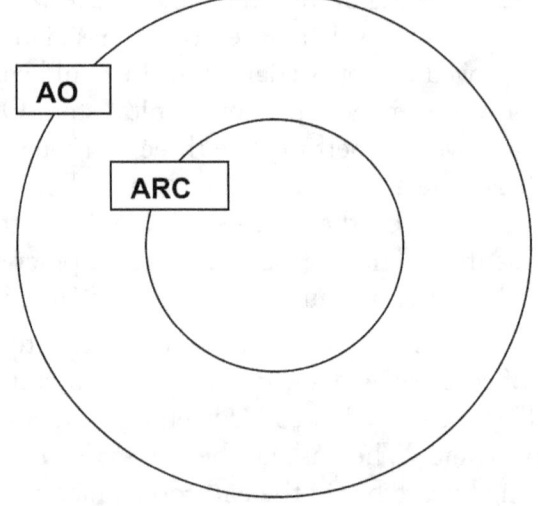

ARC in your AO

The letters ARC are an acronym for the industry-standard operational term Area of Responsibility and Control. If you were to stand up straight in an open area, let's say a parking lot or an empty garage, held a ruler in your strong hand and spun around in a giant circle with both arms fully extended out to your side – including the strong side arm with the ruler in your hand – the very outside edge of the ruler would scribe your area of responsibility and control or your ARC.

The letters AO are an acronym for the industry-standard operational term Area of Operation. If you are called out to the scene of an accident or the scene of a crime or if you are a civilian and you are out camping or hiking, or maybe a construction worker walking along the roof of a brand new commercial building, these areas as described are where you are physically working or otherwise known as area of operation or your AO.

Part III

Your Area of Responsibility and Control is at the center of your Area of Operation.

In any real world operational environment you are held fully and completely accountable for any and all activities which may occur in your ARC. If you accidentally cut yourself or someone else with your folding knife, then you are ultimately responsible as it occurred in your Area of Responsibility and Control.

The idea is to safely and effectively mange the usage of your folding knife in such a manner as to not only place your attention on what's directly in front of you, but instead focus your attention throughout your entire ARC. A less esoteric explanation is to watch what you're doing and be mindful not only of yourself but of anyone else that is within about a meter forward of your arms reach for a three-hundred and sixty degree radius of operation. It is from this perspective that we move on to building a stable working platform.

Stable Working Platform

Before we can address safe handling and usage of the folding knife we must first look at from what platform are we handling the knife.

One of the very first classes I ever taught at the US Justice Center (US FBI Academy) I was working with the instructor-trainers of the FBI's PTU (Physical Training Unit). After hours when debriefing the training I would be invited to meet in each of the trainers' offices. One common and interesting observation that caught my attention was that there was a giant pickle jar on one of the shelves of each of the offices. When I inquired "What's with the pickle jars?" the answer was quite surprising, but it hammered the point home – stable working platform.

Stable Working Platform

The explanation went something like this (plus or minus). When brand new recruits are taught to apply or remove handcuffs or something of that nature, most of these agents in training would be off balance with their feet pointing in one direction and their hands moving in a different direction and their focus (paying attention to the perceived threat) would be in an even different direction.

Noticing that their students were out of focus, hands one way and feet the other way (out of alignment), one of the instructor-trainers would command another student to run into their office and run back with the pickle jar. By the way, each of the jars had big yellow number stickers attached to each one (never asked what that was all about). The student of course came running back, out of breath, sweating and handed the pickle jar to the instructor trainer who in turn handed the pickle jar to the student who was "out of alignment." That student, holding the pickle jar, didn't really know what to make of it.

The instructor-trainer then commanded the student to "open the pickle jar" which, of course, the student immediately tried to do. It was very interesting to watch and the light bulb went off in my head as I admired the beauty and simplicity of the lesson. While the student was struggling to open the lid, I noticed that the first thing he did was pull the giant pickle jar immediately into the center of his body and brought both hands closer together. As he did this he straightened up his body and placed both feet squarely under his body weight. It was amazing to watch as the student – as a direct result of the pickle-jar technique – aligned his focus, his feet, and his hands and established a Stable Working Platform.

After returning the pickle jar, the student – now properly "aligned" was easily and effectively able to accomplish his original assignment.

In a far more domesticated example, when we make a sandwich we don't lean way over to one side and work with your hands away from your center and place your focus somewhere else (at least not most of us). No, the natural and most stable method of making a sandwich is to establish a Stable Working Platform – that is alignment of the hands, feet (weight evenly distributed over the feet) and focus of our attention throughout our area of responsibility and control.

Safe Handling

If we can easily establish a stable working platform – that is remain "aligned" with feet, hands, centered mass of the body and focused on our ARC, then we're off to a very good start in the safe handling and overall operation of the folding knife.

The two key components to safe handling of the folding knife are safely engaging the blade edge and safely engaging the blade point from a stable working platform. Let's take a closer look at each of these components – first, engaging the blade edge.

Engaging the Blade Edge

From a stable working platform there are several conventional motions that are possible with the cutting edge of the blade. First and foremost is slicing. Similar to the common action of slicing pieces of bread, the movement of pushing forward or pulling backward with the knife applying a sharpened blade edge can facilitate the cutting of various materials.

Another common action is chopping – similar to say chopping carrots on a cutting board. With the appropriate blade shape and size, chopping can be applied to a number of various materials which we will take a closer look at here soon. If the blade edge happens to be partially or fully serrated then yet another common action is made available for application of the cutting edge - sawing.

As with most evolutions of tools, the development of the blade edge was gradual and based on discoveries governed by trial and error in the field. Blade edges were a lengthy development process from the time the first hunter cleaned the first hide to this very day where modern metallurgy, molecular and particle physics plays a key role in modern edge developmental experiments.

Rigid and Flexible Cuts. Materials to which a folding knife blade may be applied may be unlimited in number, but can be easily divided into two basic categories. One is rigid or unbending materials such as hard plastics, wood, drywall, metal, etc., the other is flexible materials such as cotton, leather, light plastics, rubber, thin wire, twine, paper, cardboard, rope, nylon, etc. A folding knife edge can be effectively applied to any or all of these materials regardless of category and depending upon appropriate application of the blade edge.

Double Edge. In the world of folding knives (as was mention above) there exists no such folding knife other than a tri-fold, Balisong or stiletto (or some combination or related configuration thereof) all of which the blade must be completely secured and fully covered within the handle with no protrusions or exposed edge or tip whatsoever.

In the event that such a folding knife were to be applied, keep in mind that the tip is very thin (as a result of the double edge) which would not be at all conducive to prying and neither of the edges would be conducive to effective chopping due to the overall dagger-like shape of the blade. Suffice it to say for police work that it would be less than optimal for a peace officer to carry a folding dagger on the job.

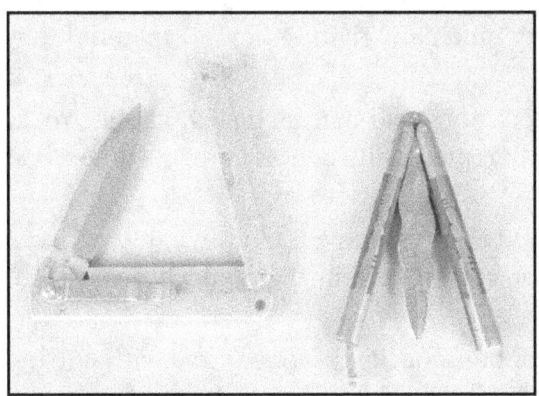

Example of folding knives with double edges (Tri-fold and Balisong).

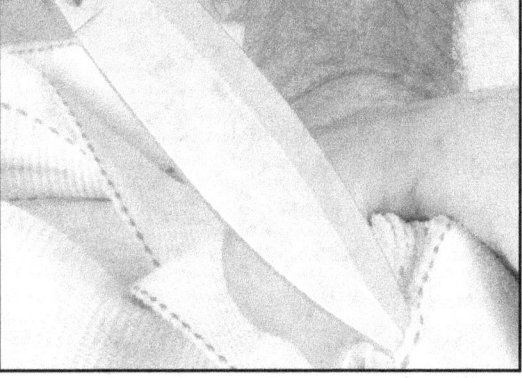

Difficulty in application of double edge.

Folding Knives: Carry & Deployment

If you are a civilian and considering carrying a folding dagger (textbook legal terminology synonymous with "double-edged folding knife") keep in mind that many state laws prohibit ownership and/or carry.

Single Edge. The majority of modern production folding knives come right out of the box with a single edge. The single edge is easily manageable, simple to open under stress and of course much easier to explain to your supervisor than a folding dagger.

The advantages of the single edge are numerous: mostly accepted by supervisors and department policies, easier (and safer) to open and close with very little skill and training (as opposed to the balisong or tri-fold), can be accessed and deployed rapidly, is not widely prohibited by varying state laws, allows for more than one method of gripping, etc., hence the reason for their popularity. One of the many advantages is that it allows a multitude of optional working grips which cannot be afforded the double edge.

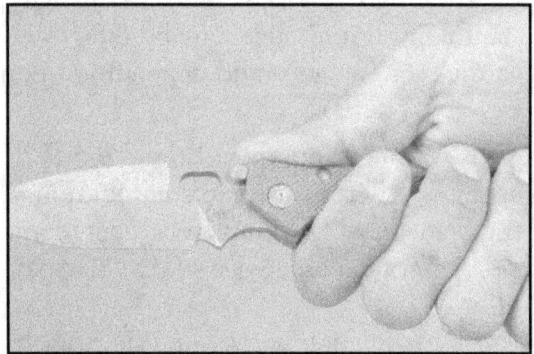 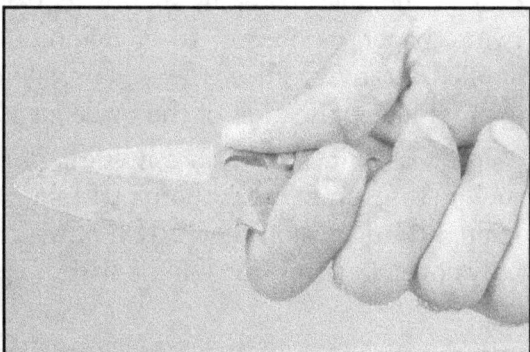

Example of an optional working grip on the single edge blade.

Aside from multiple working grips, the single edge blade also allows a two-handed Safe Open as well as a Safe Close.

Plain Edge. The plain edged has been time-tested as part of one of the oldest tools known to man. It is capable of cutting pretty much any rigid or flexible material that can be cut by a folding knife.

The plain edge can be applied to most common rigid materials such as sheetrock, ceiling tiles, cardboard, etc., as well as flexible materials such as carpet, twine, rope and webbing.

The plain edge is both versatile in usage and easy to sharpen. However, it does have some drawbacks. Sure it can make a decent cut on certain materials, but there are some limitations.

As a general observation, the straight edge has a heck of a tough time with cutting nylon rope or tubing such as seatbelts or related materials. Certain types of rope like tight composite weaves, and other cord are additionally difficult to cut with a plain edge.

Part III

The main reason for this inability to cut certain materials is the fact that the combination of tight weave, synthetic material and surface texture of the material do not allow for a deep enough "bite" of the straight-line plain edge to do the work of cutting.

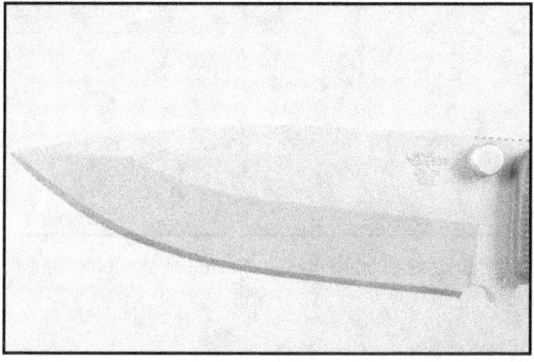

The most common blade edge type available – the Plain Edge.

Plain edge applied to braided rope.

Plain edge applied to paper.

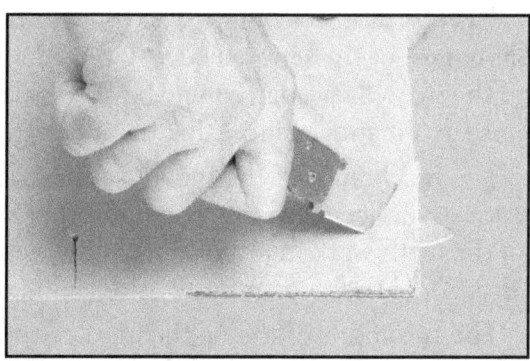

Plain edge applied to cardboard.

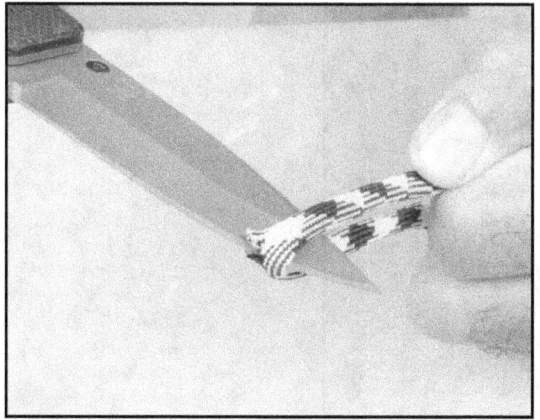

Plain edge applied to 7mm nylon weave.

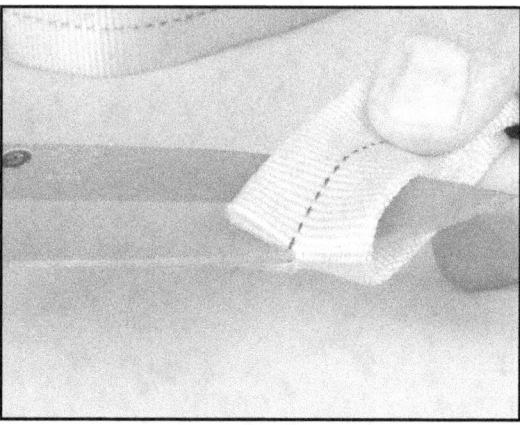

Plain edge applied to 1" nylon webbing.

137

Folding Knives: Carry & Deployment

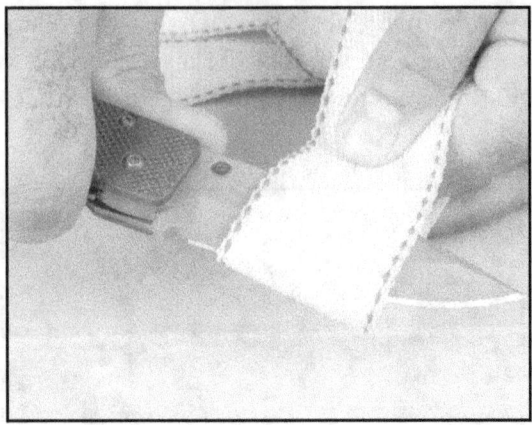
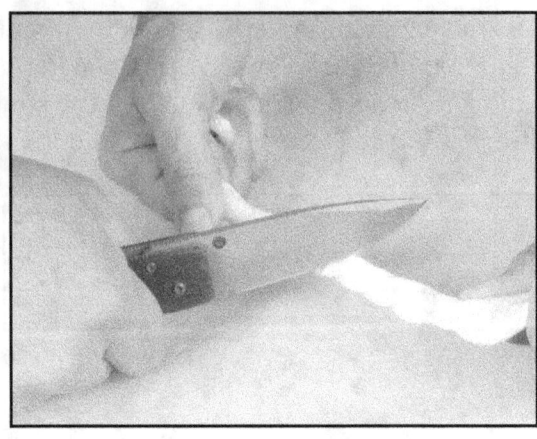

Plain edge applied to heavy duty tight-weave nylon/cotton towing strap.

Plain edge applied to double-braided nylon composite rope.

Similar to using a folding knife to saw a piece of lumber in half – OK you may be able to whittle away at it or even chop away at it, but the plain edge, as versatile as it may be, does not possess saw-like qualities.

Serrated Edge. The solution to the lack of bite and saw-like qualities problem was the invention of the serrated edge. Created by applying a multiple-ribbed grinding wheel to the steel blade edge (single bevel grind), teeth or combs are ground out of the blade in such a manner as to emulate saw teeth.

The mechanics of these combs or teeth when applied to material is a variance of ratio of surface to cutting edge. In other words the bottom of the "teeth" and the top of the teeth work in unison in adding a secondary dimension of application to the cut as opposed to the planar-only single dimension cut of the straight edge.

The net result of the application of these teeth is a saw-like capability that easily rips through the same materials with which the straight edge was struggling.

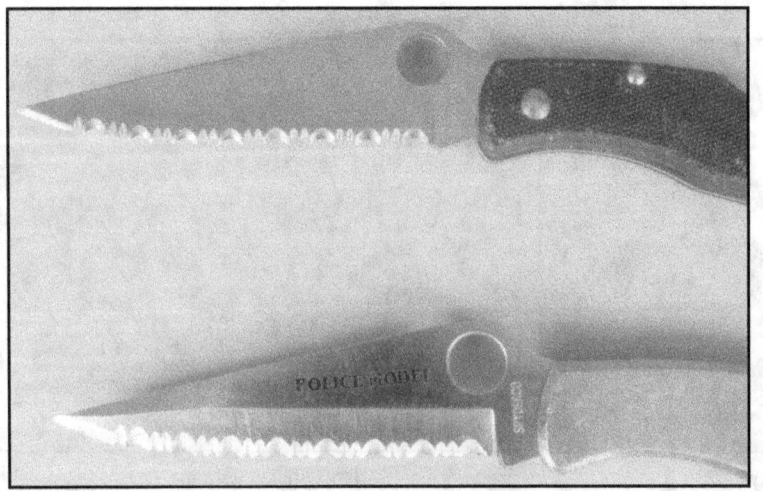

Full-serration edges.

Part III

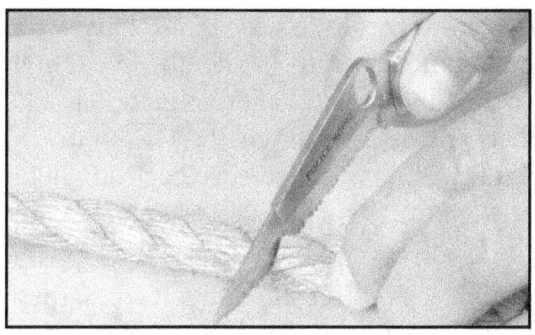
Serrated edge applied to braided rope.

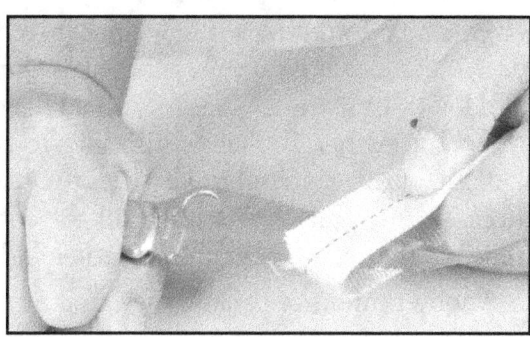
Serrated edge applied to 1" nylon webbing.

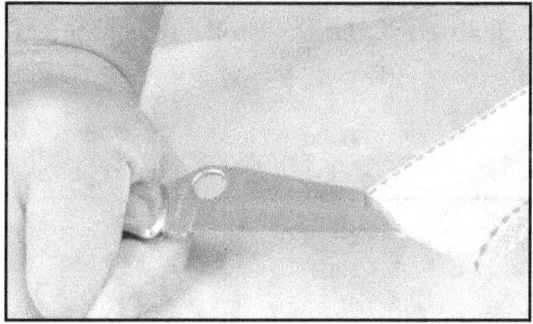
Serrated edge applied to heavy duty tight-weave nylon/cotton towing strap.

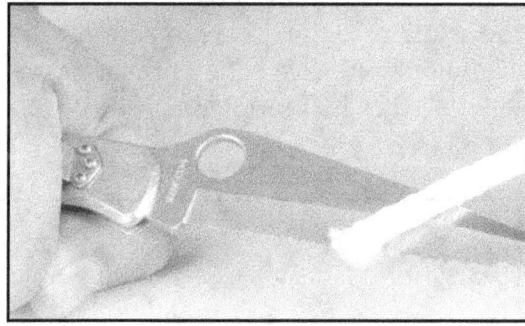
Serrated edge applied to 1" double-braided nylon composite rope.

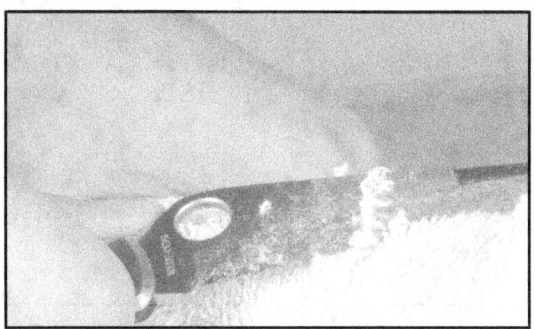
Full-serration edge applied to common bath towel.

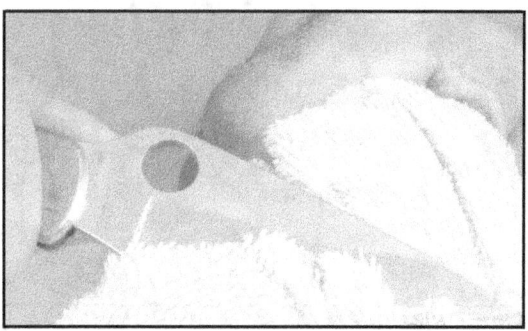
Full-serration edge applied to thick cotton sweatshirt.

Serrated edge applied to thick cardboard.

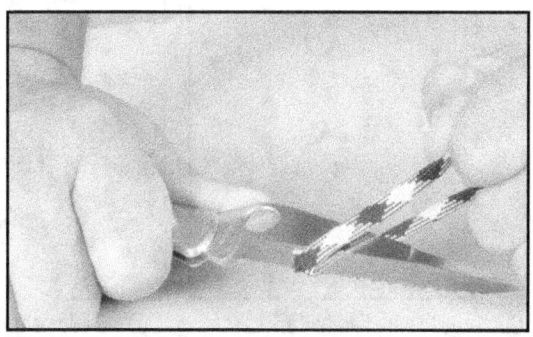
Serrated edge applied to 7mm nylon weave.

Folding Knives: Carry & Deployment

As good a solution as it may appear to be, even fully serrated edges also have some downsides and limitations. Since it is the case that saw teeth (or combs) have high and low points – much like a saw blade, if continually applied to porous materials like cotton or leather, the material begins to collect in between the teeth. The continual buildup of this material in between the teeth of the fully serrated edge begins to deteriorate the overall functionality of the cut and eventually will collect so much material that the teeth become ineffective and the blade edge will fail to cut.

Another minus of the fully serrated edge is that it is extremely difficult to sharpen. In most cases it's better to send the entire knife back to the manufacturer for professional sharpening than to try and work on one tooth or comb at a time.

Combo Edge. It didn't take much time for folding knife manufacturers to figure it out the solution to the "bite problem" of the straight edge and the "collection problem" of the serrated edge. The optimal answer was the "combo edge" or what some folks may call the "half-serrated edge."

Combination Edge.

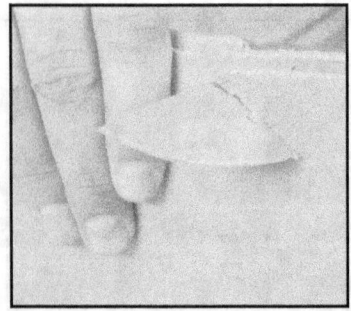

Combo edge applied to thick cardboard.

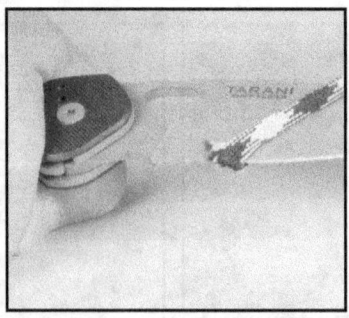

Combo edge applied to 7mm nylon weave.

Combo edge applied to braided rope.

Part III

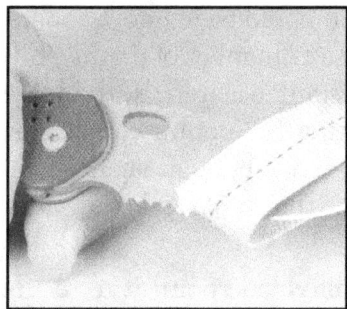
Combo edge applied to 1" nylon webbing.

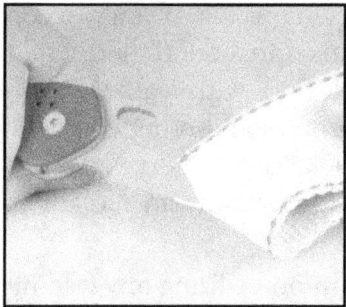
Combo edge applied to heavy duty tight-weave nylon/ cotton towing strap.

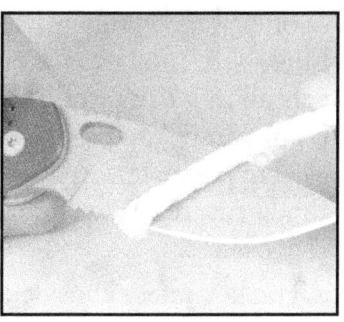
Combo edge applied to 1" double-braided nylon composite rope.

The combo edge, with regards to a folding knife blade, offers the best of both worlds. Where the plain edge part of the blade fails the serrated edge part of the blade follows immediately behind to take up the slack and vice versa.

As useful as the combo-edge is in combining the best of both worlds, the converse is also true. Same as the other two edges, the combination edge also has its downside. Certain materials are cut more effectively with a full plain edge while others are cut more effectively with a fully serrated edge.

A final word on plain edges, fully serrated edges and combination edges is yet again – application. What exactly are you planning on using the edge for? Each of these three edge grind configurations has a plus and a minus. Specialized application of any edge will determine an edge type. For example, if your job requires you to cut sheetrock all day, then your best bet is going to be a sheetrock knife specifically designed to cut sheetrock. Same thing goes for a box-cutter or a carpet-cutter. However, for generalized utility application any of these three edges are well-designed to effectively handle the job.

Engaging the Point

The second key component to safe handling of the folding knife is safely engaging the blade point from a Stable Working Platform. The point, again for utility purpos-

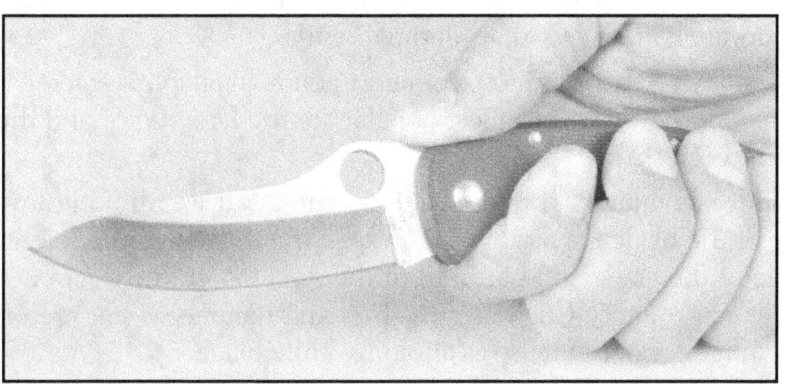
Utilizing the point in a thrusting motion with an optimal grip.

Folding Knives: Carry & Deployment

es only, is predominantly used for prying. If you're a cop this could be windows, doors, hinges, etc., and if you work in construction there could be any number of prying projects all in a day's work. If you work with leather or other similar materials, the knife point can be used for poking holes and otherwise perforating. The point can also be used to pierce screen doors, poke through drywall and if it's good enough quality steel (and heat treat) a well-made knife point can actually pierce the trunk or door of a car if needed.

The key safety consideration on utilizing any folding knife blade point is to secure a very firm and stable positive grip. Utilizing the friction radius and the overall design of the handle, it is a matter of operational safety to maintain a positive and secure grip on the folding knife handle at all times during usage of the point.

When engaging the point, it is critical to maintain a stronger-than-usual grip on the knife handle and maybe, given certain models, even on part of the back of the blade. The price for not having a solid grip and to properly utilize the friction radius is that once the point makes contact with whatever material, there is a tremendous amount

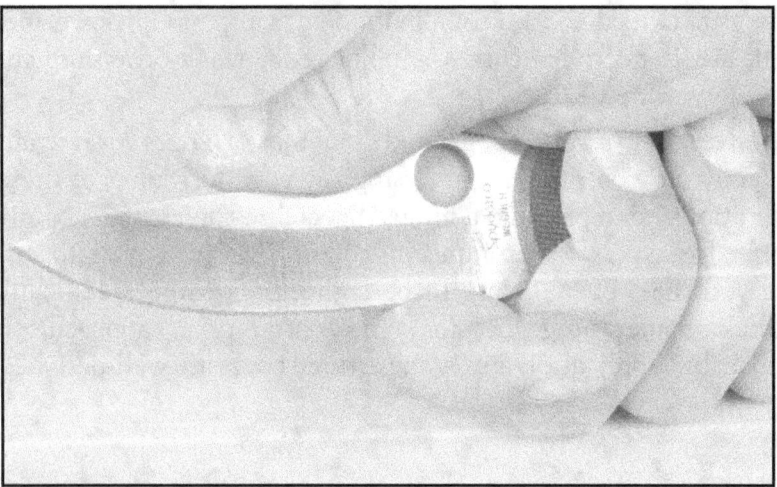

Utilizing the point in a thrusting motion with a less-than-optimal grip.

of pressure applied to the "opposing force vectors" (as per above) and what you end up with is tremendous strain on the surface contact between the hands, fingers and the blade handle. Should your grip be pried loose from the handle, force will continue to drive your hand forward toward less-than-optimal results.

Limiting our scope of study to the two most common points available, remember there is an endless number of these, the two most popular are the Drop Point and the Angle Point (Tanto or Yoroi Toshi).

Drop Point. Although not recommended by the manufacturer, but we all do it anyway, is to utilize the tip of the blade for prying. The reason the manufacturer doesn't recommend it is because of the grade and hardness of the steel. Most knives are not manufactured for "tool-grade" quality (such as crow-bars and hammers), but are on the other hand manufactured for optimal use as a folding knife blade.

Regarding thrusting, the design of the drop point utilizes optimal blade geometry (as was presented earlier in Metallurgy and Blade Geometry) in such a manner as to facilitate the dispersal of forward energy in an "angular momentum" which optimizes (acting as a force vector camming device) thrust. In knuckle dragger terminology, it's a great design for short thrusting motions.

Upon piercing material, the width of the blade increases which eventually increases the amount of resistance on the material and decreases the thrusting energy. Generally speaking the Drop Point is sufficient for most utility application.

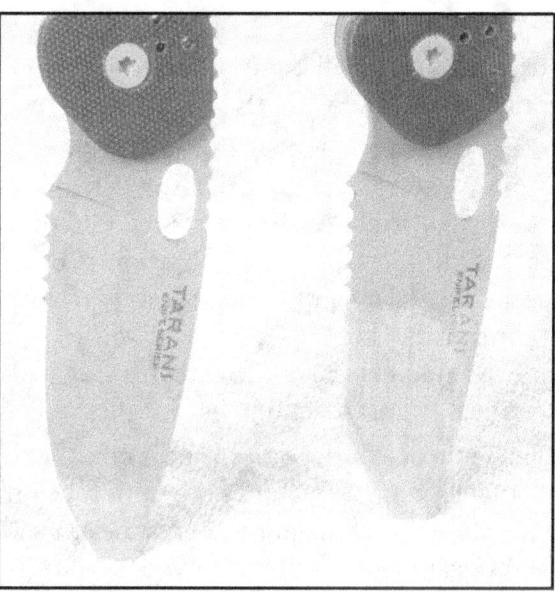
Difference between Drop Point and Angle Point applied to rigid material.

Angle Point. As mentioned above (see Blade Points) Angle Point or Tanto style goes back literally centuries and can be traced to Asian origins. It is classically assigned to the Japanese blade heritage which began with the highly skilled blade-makers of antiquity who forged the early renditions of the Katana and the Wakazashi.

How did this particular point design gain its reputation as an efficient thrusting tool? As mentioned above, in ancient times it was nick-named "Yoroi Toshi" or "body armor piercing" a term that stuck around (no pun intended).

The design features of the Tanto are simple, it's a thick stout blade tip with few edge bevels. Mechanically speak-

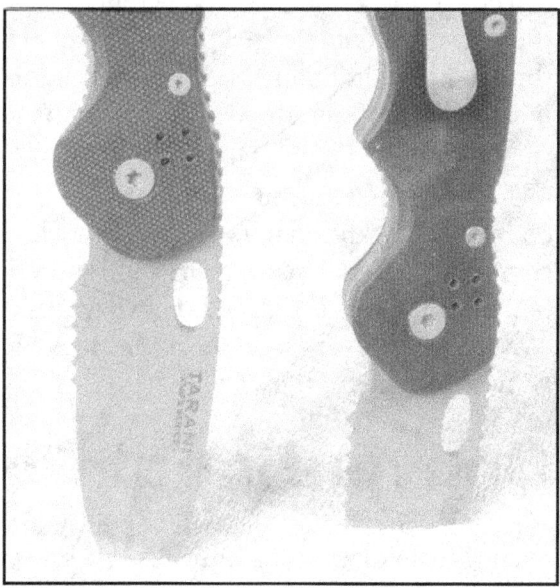
Performance of Drop Point and Angle Point when equal force is applied to each against rigid material.

ing, the stout tip allows for optimal prying capability (as there is a lot more steel concentrated in the small area of the point compared to other blade point geometry). Additionally the overall blade design is relatively uniform – basically the same thickness of the blade all the way to the base of the blade, which contributes to a minimal coefficient of friction (we'll look more in depth at this friction stuff in this next section) allowing more depth in penetration than the drop point design.

Coefficient of Friction. In all blade designs there are two major factors which contribute to the coefficient of friction (amount of drag) generated during a chop, thrust or edge cut. These are the number of bevels and blade geometry.

Typically a single-bevel or chisel-ground edge (say a "Tanto"-style blade) demonstrates less friction on chopping through rope or thrusting through rugged material than a double bevel (say a double-edge double-bevel spear point.) The determining factor of this coefficient of friction is both the edge geometry (the number of bevel surfaces) and overall blade geometry (shape of the blade – spear point, drop point, clip point, etc.) that make contact with the material. The more contact against the greater number of surfaces creates the greatest amount of friction (or drag). The less contact against less number of surfaces reduces drag.

The tip of a drop point or spear point, for example, begins with minimal surface area contact with material (as penetration begins at the tip) but increases exponentially as the blade, by its geometry, continues to widen as it moves forward through the material toward the handle. The widening of the blade causes additional surface contact which increases the coefficient of friction.

A Tanto point chisel-ground blade with its single-bevel edge has one half the drag of a conventional double-bevel ground blade. It also has between a quarter and a sixth the drag of a double-edged double-bevel blade as a result of less surface contact when thrusting. Additionally the Tanto point blade geometry (overall width) does not increase in size from the base of the point to the handle during insertion which additionally contributes to a decrease coefficient of friction.

During application of the point, if a blade maintains a consistent multiple bevel contact and additionally increase in width, this will increase the coefficient of friction. If a blade is consistent in width and sustains minimal surface contact, the coefficient of friction is reduced. However, the advantages and disadvantages of number of bevels and blade geometry, as one may by now expect, is determined by the intended functionality of the blade.

Conventional Positions

The final stages of training in the safe and effective handling of a folding knife include utilizing the folding knife in varying body positions. There are two categories of body positioning and these are conventional and unconventional positions. The more common of the two categories is the conventional positions category which includes the standing, kneeling and seated positions.

Standing

The first of these (and believe it or not least stable) is the standing position. We as humans are bi-pods and when standing have only two points of contact with the ground. Whereas a horse, dog or cat has four points of contact with the ground.

Part III

In order to safely and effectively operate the knife from the standing position it is first necessary to establish a stable working platform. If we recall from above, a Stable Working Platform is the alignment of the hands, feet (weight evenly distributed over the feet) and focus of our attention throughout our area of responsibility and control.

Utilizing the Stable Working Platform it is simply a matter of applying the knife to whatever material in such a manner as to maintain the stability of the working platform.

Obviously there will be times where you may want to take a step or adjust your position. It is strongly recommended to not move abruptly while the blade is in the open and locked position in the event that your footing is unsure (terrain, inclement weather, viscous material under your boots, etc.) or that you may be in or on a moving vehicle.

Kneeling

Another common or conventional position is the kneeling position. It is the next level of elevation toward the ground below the standing position.

Utilizing the Stable Working Platform in the Standing Position. Notice the alignment of center mass, feet shoulder-width or greater apart and hands near or at center for optimal application of core strength in control of the knife blade.

Kneeling

A less-than-optimal placement of the knees ("surfing") in the kneeling position can cause instability (loss of the Stable Working Platform) which may cause potentially injurious repercussions especially fumbling with an open and locked knife blade.

Much like a camera tripod the minimal amount of contact in the kneeling position is three points of solid contact with the ground. Notice wide stable base.

In this example more than three points of contact are applied to the ground with both knees forming an optimal Stable Working Platform

If you asked an experienced rock climber how many points of contact is optimal for climbing – he/she would most likely respond "as many points as possible." In my younger days I used to climb and can remember times where I wished I had even more body parts. The moral of the story is, the more contact points you have with the rock the more stable your position. This coincides with our earlier chapters on grips of the folding knife where it was discovered that "more contact equals more control." Same applies here.

Part III

Optimally, to establish a stable working platform in the kneeling position it's important to establish three solid points of contact with the ground.

Seated

The next elevation down from kneeling moving closer to the ground is one of the most common (and, of course, comfortable) of the conventional positions is seated.

Checking off all the boxes: At least three points of contact with the ground, alignment of center mass and position of the hands close in to the center of the body and operating the knife from a Stable Working Platform, the seated position is the most stable and comfortable of the conventional positions.

Stability and comfort are important if it is necessary to work in the same position for long periods of time – perhaps as an electrician or plumber.

Seated

Example of safely utilizing the folding knife from a Stable Working Platform in the seated position.

Unconventional Positions

At the opposite end of the spectrum from conventional body positions are unconventional body positions. First responders are probably the most familiar with these positions. Often times at the scene of an accident or natural disaster it's necessary to rapidly and effectively apply the folding knife from some awkward or unconventional bodily position. Linemen, construction workers, plumbers, electricians, and handymen are also no strangers to the unconventional position with regards to using a knife.

The same rules that apply to the conventional body positions directly apply to the unconventional positions. As mentioned above the critical handling points are to establish a stable platform with at least three points of contact on the ground and be able to use the knife from this position in a safe and controlled manner.

Prone

An often-seen unconventional position is the prone position. Facing the ground, the prone position is where you would need to move your body to the lowest elevation toward the ground (one step lower even than sitting). There are a couple of ways to configure your body parts in this prone position to facilitate a stable working platform and utilizing the knife in a safe and controlled manner from an awkward position. The first of these is with legs evenly spaced and head straight up.

Establishing a Stable Working Platform from the prone position with legs spaced evenly apart – military prone.

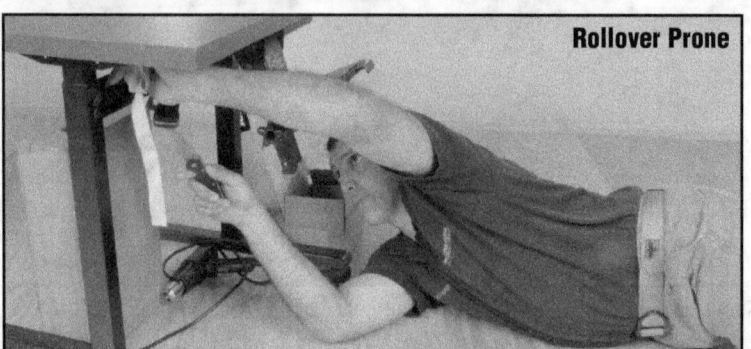

Establishing a Stable Working Platform from the prone position with the need to be raised slightly off the ground and on angle – rollover prone.

Part III

Supine

The opposite of the prone is the supine position. Every auto mechanic in the world is familiar with this position. It's sometimes required to work with a knife from a position where you must lay flat on your back and looking straight up at your work area. Similar to the prone position, there are a couple different versions of this, but the most common of the two is to place your entire body (except for the hands of course) flat on the ground. In all honesty this is the most stable and of course most comfortable of any of the positions (both conventional and unconventional).

One of the only safety checks you may need to run on this is to keep your work area far enough away from your face that if a piece of material or even the knife came flying away from your hands that it would miss your face completely.

Example of safely utilizing the folding knife from a Stable Working Platform in the supine position.

Asymmetric

In the real world not all bodily positions can be neatly covered as per above, what about the way off-hand unconventional position that is "none of the above?" The most unconventional of the unconventional positions is often times referred to as "Asymmetric."

The term asymmetric literally means "not symmetrical" which, in terms of body positions means that one part is placed in such a way over here while another part or parts are placed in a different location. The net result is you end up needing to work with your folding knife in an awkward, uncomfortable and unconventional body position.

Folding Knives: Carry & Deployment

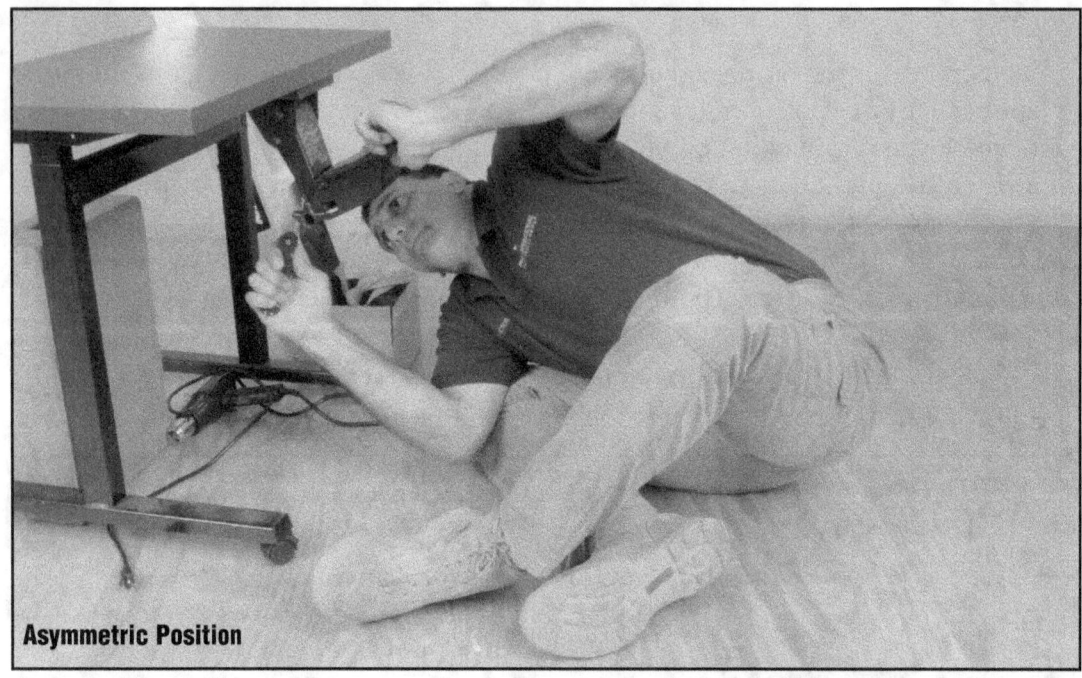

Asymmetric Position

Example of safely utilizing the folding knife from a Stable Working Platform in an awkward position.

In such an awkward body position the steps are the same as all of the above – step 1. Establish a Stable Working Platform, Step 2. I not in the standing position then establish three solid points of stable contact with the ground. Step 3. Utilize the knife in a safe and controlled manner from this Stable Working Platform.

• •

Asymmetric Position - ALT

Example of safely utilizing the folding knife from a Stable Working Platform in alternate asymmetric position.

Part III

Conclusion

As previously covered, folding knives, as a problem solving tool, are an important piece of the modern utility kit. In these previous chapters a tremendous amount of information about folding knives was presented: History, Categories, Metallurgy, Geometry, Parts, Blade Shape, Points, Edges, Grinds, Finishes, Handles, Opening Mechanisms, Locking Mechanisms, Selection, Maintenance, Accessibility, Carry, Locking, Unlocking, Grips, Deployment, Rapid Deployment, Position, Balance, Safe Handling, Conventional and Unconventional Positions to name a few.

It is the intention of this author to pass on as much valuable information as possible to assist in the selection, carry and usage of the modern folding knife.

As a full-time professional instructor, I share something with my students that my teachers shared with me many years ago. It is customary for a student after completing any training to ask themselves two important personal questions at the end of every training program - the same applies to the material in this manuscript: "Did I learn something new?" and "Did I learn something useful?"

Thanks for taking the time to read through this detailed material. I certainly hope that this information proves useful and the book an enjoyable read.

If you may be looking for professional programs of instruction on the selection, carry and usage of the modern folding knife for your department or agency please contact the staff at Operational Skills Group, LLC at *info@opskillsgroup.com*. If you may have any questions, comments, concerns regarding this material, I can be reached via the Operational Skills Group, LLC at *steve.tarani@opskillsgroup.com* when available.

Stay safe and stay trained!

Steve Tarani

About the Author

Steve Tarani is an internationally respected authority within the professional law-enforcement and military training community. He is a full-time training consultant providing high-profile operational skills development programs for various US agencies.

Tarani is unparalleled in the law enforcement training community as a master-level instructor and subject matter expert (SME) specializing in use-of-force training. Steve is also the developer of well known instructor certification programs which have gained both federal and Peace Officer Standards of Training accreditation.

Author delivering lecture on Edged Tools.

Brief background:

- On instructor staff at the U.S. D.O.E. Nonproliferation and National Security Institute (Central Training Academy) Security Force Training Dept. at Kirtland Air Force Base (NM).
- Defensive tactics advisor to US Federal Bureau of Investigation (FBI), US Transportation Security Administration (ISD ACSP) and the U.S. State Dept. Bureau of Diplomatic Security Anti-terrorism Assistance Program. (ATAP)
- Sworn in the state of California serving on staff as Senior Defensive Tactics and Firearms Instructor for Del Rey Oaks Police Department in reserve capacity.
- On instructor staff as a Firearms Rangemaster at Gunsite Academy (AZ).
- CEO of Operational Skills Group, LLC as a full-time educator and training management consultant throughout the professional law enforcement, specialized military, law enforcement and US federal training communities.
- Sworn in the State of Nevada as a Deputy for Pershing County Sheriff's Office.
- Federally certified firearms instructor (US DOE, US DoD, etc.).
- Uniquely trained in documented edged tool disciplines, certified in three specific systems of instruction in the US, Philippines and Indonesia.

NOTES

NOTES

NOTES

NOTES

NOTES

NOTES

NOTES

NOTES

www.ingramcontent.com/pod-product-compliance
Lightning Source LLC
Chambersburg PA
CBHW080025130526
44591CB00037B/2673